Correlation Clustering

Synthesis Lectures on Data Mining and Knowledge Discovery

Editors

Jiawei Han, *University of Illinois at Urbana-Champaign*
Johannes Gehrke, *Cornell University*
Lise Getoor, *University of California, Santa Cruz*
Robert Grossman, *University of Chicago*
Wei Wang, *University of North Carolina, Chapel Hill*

Synthesis Lectures on Data Mining and Knowledge Discovery is edited by Jiawei Han, Lise Getoor, Wei Wang, Johannes Gehrke, and Robert Grossman. The series publishes 50- to 150-page publications on topics pertaining to data mining, web mining, text mining, and knowledge discovery, including tutorials and case studies. Potential topics include: data mining algorithms, innovative data mining applications, data mining systems, mining text, web and semi-structured data, high performance and parallel/distributed data mining, data mining standards, data mining and knowledge discovery framework and process, data mining foundations, mining data streams and sensor data, mining multi-media data, mining social networks and graph data, mining spatial and temporal data, pre-processing and post-processing in data mining, robust and scalable statistical methods, security, privacy, and adversarial data mining, visual data mining, visual analytics, and data visualization.

Correlation Clustering
Francesco Bonchi, David García-Soriano, and Francesco Gullo
2022

Detecting Fake News on Social Media
Kai Shu and Huan Liu
2019

Multidimensional Mining of Massive Text Data
Chao Zhang and Jiawei Han
2019

Exploiting the Power of Group Differences: Using Patterns to Solve Data Analysis Problems
Guozhu Dong
2019

Mining Structures of Factual Knowledge from Text
Xiang Ren and Jiawei Han
2018

Individual and Collective Graph Mining: Principles, Algorithms, and Applications
Danai Koutra and Christos Faloutsos
2017

Phrase Mining from Massive Text and Its Applications
Jialu Liu, Jingbo Shang, and Jiawei Han
2017

Exploratory Causal Analysis with Time Series Data
James M. McCracken
2016

Mining Human Mobility in Location-Based Social Networks
Huiji Gao and Huan Liu
2015

Mining Latent Entity Structures
Chi Wang and Jiawei Han
2015

Probabilistic Approaches to Recommendations
Nicola Barbieri, Giuseppe Manco, and Ettore Ritacco
2014

Outlier Detection for Temporal Data
Manish Gupta, Jing Gao, Charu Aggarwal, and Jiawei Han
2014

Provenance Data in Social Media
Geoffrey Barbier, Zhuo Feng, Pritam Gundecha, and Huan Liu
2013

Graph Mining: Laws, Tools, and Case Studies
D. Chakrabarti and C. Faloutsos
2012

Mining Heterogeneous Information Networks: Principles and Methodologies
Yizhou Sun and Jiawei Han
2012

Privacy in Social Networks
Elena Zheleva, Evimaria Terzi, and Lise Getoor
2012

Community Detection and Mining in Social Media
Lei Tang and Huan Liu
2010

Ensemble Methods in Data Mining: Improving Accuracy Through Combining Predictions
Giovanni Seni and John F. Elder
2010

Modeling and Data Mining in Blogosphere
Nitin Agarwal and Huan Liu
2009

Correlation Clustering

Francesco Bonchi, David García-Soriano, and Francesco Gullo

ISBN: 978-3-031-79198-7   paperback
ISBN: 978-3-031-79210-6   PDF
ISBN: 978-3-031-79222-9   hardcover

DOI 10.1007/978-3-031-79210-6

A Publication in the Springer series
SYNTHESIS LECTURES ON DATA MINING AND KNOWLEDGE DISCOVERY

Lecture #19
Series Editors: Jiawei Han, *University of Illinois at Urbana-Champaign*
                Johannes Gehrke, *Cornell University*
                Lise Getoor, *University of California, Santa Cruz*
                Robert Grossman, *University of Chicago*
                Wei Wang, *University of North Carolina, Chapel Hill*
Series ISSN
Print 2151-0067   Electronic 2151-0075

Correlation Clustering

Francesco Bonchi
ISI Foundation, Turin, Italy

David García-Soriano
ISI Foundation, Turin, Italy

Francesco Gullo
UniCredit, Rome, Italy

SYNTHESIS LECTURES ON DATA MINING AND KNOWLEDGE DISCOVERY #19

ABSTRACT

Given a set of objects and a pairwise similarity measure between them, the goal of correlation clustering is to partition the objects in a set of clusters to maximize the similarity of the objects within the same cluster and minimize the similarity of the objects in different clusters. In most of the variants of correlation clustering, the number of clusters is not a given parameter; instead, the optimal number of clusters is automatically determined.

Correlation clustering is perhaps the most natural formulation of clustering: as it just needs a definition of similarity, its broad generality makes it applicable to a wide range of problems in different contexts, and, particularly, makes it naturally suitable to clustering structured objects for which feature vectors can be difficult to obtain. Despite its simplicity, generality, and wide applicability, correlation clustering has so far received much more attention from an algorithmic-theory perspective than from the data-mining community. The goal of this lecture is to show how correlation clustering can be a powerful addition to the toolkit of a data-mining researcher and practitioner, and to encourage further research in the area.

KEYWORDS

clustering, algorithmic theory, computational complexity, approximation algorithms, graph partitioning, agnostic learning, active learning, consensus clustering, clustering aggregation, constrained clustering, local clustering, overlapping clustering, graph coloring, dimensionality reduction, latent features learning, edge-labeled graphs, multilayer networks, document classification, duplicate detection, conversation disentanglement, image segmentation, community detection

Contents

Preface xiii

Acknowledgments xv

1   Foundations 1

   1.1   Motivation and Basic Formulation 1

   1.2   Early Work on Correlation Clustering 2

   1.3   Computational Hardness 3

   1.4   A More General Formulation 5

       1.4.1   Max-Agreement Correlation Clustering 7

   1.5   Approximation Algorithms for MIN-DISAGREE 8

       1.5.1   Approximation Algorithms for Complete Graphs 8

       1.5.2   Extension to Probability Constraints 11

       1.5.3   An $O(\log n)$-Approximation for General Weighted Graphs 13

   1.6   Approximation Algorithms for MAX-AGREE 15

   1.7   Correlation Clustering in Presence of a Ground Truth 15

   1.8   Related Problems 17

   1.9   Applications 20

2   Constraints 23

   2.1   Correlation Clustering with a Fixed Number of Clusters 23

       2.1.1   Hardness Results 24

       2.1.2   Polynomial-Time Approximation Schemes 26

       2.1.3   Other Approximation Results and Algorithms 30

       2.1.4   General Graphs 32

   2.2   Correlation Clustering with Constrained Cluster Sizes 34

       2.2.1   A Region-Growing Algorithm 36

       2.2.2   A (Randomized) Pivoting Algorithm 40

       2.2.3   Non-Uniform Constrained Cluster Sizes 42

   2.3   Correlation Clustering with Error Bounds 42

       2.3.1   Connection with BOUNDED-MIN-MULTICUT 43

       2.3.2   An Approximation Algorithm 45

3 Relaxed Formulations **51**

3.1 Overlapping Correlation Clustering 51

    3.1.1 Constraints and Characterization 52

    3.1.2 Hardness Results 53

    3.1.3 Connection with Graph Coloring 54

    3.1.4 Connection with Intersection Representation 55

    3.1.5 Connection with Dimensionality Reduction 56

    3.1.6 Connection with Nonnegative Matrix Factorization ... 56

    3.1.7 Connection with Latent Features Learning 57

    3.1.8 Algorithms 58

3.2 Correlation Clustering with Local Objectives 61

    3.2.1 Vertex-Wise Formulations 63

    3.2.2 Cluster-Wise Formulation 74

3.3 Correlation Clustering with Outliers 74

4 Other Types of Graphs **77**

4.1 Bipartite Graphs 77

    4.1.1 Pivot-BiCluster Algorithm 79

4.2 Edge-Labeled Graphs 80

    4.2.1 Approximation Algorithms 82

    4.2.2 Heuristics 87

4.3 Multilayer Graphs 90

    4.3.1 Algorithms 92

4.4 Vertex-Labeled Graphs 94

4.5 Hypergraphs 95

4.6 Noisy Graphs 97

5 Other Computational Settings **99**

5.1 Query-Efficient Correlation Clustering 99

    5.1.1 Analysis of QECC 100

    5.1.2 A Non-Adaptive Algorithm 106

    5.1.3 A Practical Improvement 107

    5.1.4 Neighborhood Queries 108

5.2 Local Correlation Clustering 109

5.3 Large-Scale Computing 110

    5.3.1 Online Setting 110

    5.3.2 Parallel Computing 110

    5.3.3 Dynamic Data-Stream Model 111

6   **Conclusions and Open Problems** 113

   Bibliography 115

   Authors' Biographies 131

Preface

Clustering is one of the most well-studied problems in data mining. The goal of clustering is to group a set of objects so that objects in the same cluster are more similar to each other than to objects in other clusters. When formulated as an optimization problem, this abstract description of a clustering task gives naturally rise to a problem known as *correlation clustering*. Given a set of objects and a pairwise similarity measure between them, the goal of correlation clustering is to partition the objects in a set of clusters so as to maximize the similarity of the objects within the same cluster and minimize the similarity of the objects in different clusters. In most of the variants of correlation clustering, the number of clusters is not a given parameter; instead, the optimal number of clusters is automatically determined.

Correlation clustering is perhaps the most natural formulation of clustering: as it just needs a definition of similarity, its broad generality makes it applicable to a wide range of problems in different contexts, and, particularly, makes it naturally suitable to clustering structured objects for which feature vectors can be difficult to obtain. Despite its simplicity, generality, and wide applicability, correlation clustering has so far received much more attention from an algorithmic-theory perspective than from the data-mining community. The goal of this book is to show how correlation clustering can be a powerful addition to the toolkit of a data-mining researcher and practitioner, and to encourage further research in the area.

The main focus of this book is on the basic problem definitions, the algorithms, and the main theoretical results for correlation clustering. The key algorithms are provided and their associated analysis is presented in detail. For a selection of the most important theoretical results, their proofs are also provided. These fundamental notions are also collocated within a wide literature of related problems and techniques. The lecture also provides a thorough overview of a series of variants of the basic correlation-clustering problem, with an emphasis on the algorithmic techniques and key ideas developed to derive efficient solutions. Finally, the lecture also discusses real-world applications, scalability issues that may arise, and the existing approaches to handle them.

The lecture is organized as follows. Chapter 1 presents the foundations of correlation clustering, the basic problem formulations and the main algorithmic results.

Chapter 2 discusses alternative formulations of correlation clustering where *constraints* are added to the basic formulation: these include formulations of correlation clustering where the number of output clusters is fixed, where the size of the output clusters is bounded by some constant(s), and where the output clustering has a pre-specified error.

Chapter 3 deals with *relaxed* formulations of correlation clustering, i.e., formulations where some constraints of the basic formulation are discarded or required in a less restrictive form: these includes correlation clustering where the output clusters are allowed to overlap, correlation clustering where (dis)agreements are considered locally, at the level of a single vertex, and the goal is to optimize an aggregation function of the local (dis)agreements, and correlation clustering with outliers, i.e., when not necessarily all the objects need to be clustered and some of them might be discarded.

Chapter 4 presents variants of correlation clustering on different types of graphs, e.g., bipartite graphs, edge-labeled and vertex-labeled graphs, multilayer graphs, hypergraphs.

While previous chapters analyze the algorithms in the standard memory RAM model, Chapter 5 discusses correlation clustering in computational models that go beyond the traditional one, e.g., large-scale, parallel, streaming, online settings.

Finally, Chapter 6 concludes the lecture with a discussion of the most important open problems.

The book is intended for students, researchers, and practitioners in Computer Science and Data Science with an interest in algorithmic theory. It is written so as to be accessible to anyone familiar with the general area of algorithms at the level of a standard undegraduate course [Cormen et al., 2009, Kleinberg and Tardos, 2006, Manber, 1989]. Notions of complexity theory and the theory of NP-completeness [Garey and Johnson, 1979], as well as basic knowledge of approximation algorithms and the use of linear-programming relaxations (covered in, e.g., Chapters 1–20 of Vazirani [2001] or Chapters 1–7 of Williamson and Shmoys [2011]) are useful prerequisites, although not strictly needed.

Francesco Bonchi, David García-Soriano, and Francesco Gullo
January 2022

Acknowledgments

We wish to thank our friends and collaborators Aristides (Aris) Gionis, Edo Liberty, Domenico Mandaglio, Andrea Tagarelli, Charalampos (Babis) Tsourakakis, and Antii Ukkonen for many fruitful discussions about correlation clustering. Edo Liberty is one of the co-authors of our KDD 2014 tutorial [Bonchi et al., 2014a], which is the starting seed for this lecture.

Francesco Bonchi and David García-Soriano acknowledge support from Intesa Sanpaolo Innovation Center. The funder had no role in study design, data collection and analysis, decision to publish, or preparation of the manuscript.

Francesco Bonchi, David García-Soriano, and Francesco Gullo
January 2022

CHAPTER 1

Foundations

1.1 MOTIVATION AND BASIC FORMULATION

Clustering is the process of grouping a set of objects into "clusters" so that objects in the same cluster are more similar (in some sense) to each other than to those in other clusters. It is one of the most extensively performed tasks in exploratory data mining, machine learning and pattern recognition. *Correlation clustering* (or *cluster editing*) is arguably the most natural formulation of clustering. In the simplest setting, we are given a set V of objects and a symmetric pairwise similarity function $s : \binom{V}{2} \to \{0, 1\}$, where $\binom{V}{2}$ is the set of unordered pairs of objects in V. The similarity function $s(u, v)$ equals 1 when u and v are similar and hence should be clustered together, and 0 otherwise. The goal is to cluster the objects in such a way that, to the best possible extent, similar objects are put in the same cluster and dissimilar objects are put in different clusters.

A *clustering* of V is a partition of V, i.e., a decomposition of V into non-empty disjoint sets $V_1, \ldots, V_k$ whose union is V. We can represent any clustering of V by a labeling function $\ell : V \to \mathbb{N}$ which gives integer cluster identifiers for each element of V; each cluster is a maximal set of vertices sharing the same label. Correlation clustering, as defined below, aims to minimize the following cost.

> **Problem 1.1 (Correlation-Clustering)** Given a set V of objects and a pairwise similarity function $s : \binom{V}{2} \to \{0, 1\}$ defined on unordered pairs of elements of V, find a clustering $\ell : V \to \mathbb{N}$ that minimizes the following cost:
>
> $$\text{cost}(\ell) = \sum_{\substack{(x,y) \in \binom{V}{2}, \\ \ell(x) = \ell(y)}} (1 - s(x, y)) + \sum_{\substack{(x,y) \in \binom{V}{2}, \\ \ell(x) \neq \ell(y)}} s(x, y). \qquad (1.1)$$

The intuition underlying the above problem definition is that if two objects x and y are dissimilar and are assigned to the same cluster, one pays a cost of 1 for their dissimilarity. Similarly, if x, y are similar and they are assigned to different clusters, one pays cost 1 as well, for their similarity $s(x, y)$. In a way, we are seeking the clustering that correlates best with the given similarity function.

The binary-similarity setting can be viewed very conveniently through graph-theoretic lenses: the objects in V correspond to the vertices of a *similarity graph* $G = (V, E)$, which is an undirected graph where an edge is placed between vertices u and v whenever $s(u, v) = 1$, i.e., $E = \{\{x, y\} \in \binom{V}{2} \mid s(x, y) = 1\}$. We can represent a clustering ℓ with a *cluster graph* C_ℓ, i.e., a union of vertex-disjoint cliques: each clique of C_ℓ represents a cluster of C. If G is a cluster graph by itself, then its connected components give a zero-cost clustering of V. However, in general, some similarities will be inconsistent with one another because similarities may not be transitive, so incurring a certain cost is unavoidable: this is the case, for example, when $s(x, y) = 1$ and $s(y, z) = 1$ but $s(x, z) = 0$. If we define the *edit distance* between two graphs $G = (V, E)$ and $H = (V, E')$ by $\mathrm{dist}(G, H) = |E \oplus E'|$, Problem 1.1 asks for the cluster graph H with minimum edit distance from the input graph G. The problem may also be viewed as the task of finding the equivalence relation that most closely resembles a given symmetric relation.

The framework above naturally extends to non-binary, symmetric similarity functions, and to settings where the similarity information is incomplete, i.e., no penalty needs to be paid for some pairs of objects irrespective of whether they are placed together or not (Section 1.4).

A key feature of correlation clustering is that it does not require the number of clusters to be given as input, in contrast to many other popular clustering techniques (such as k-means); instead, it automatically finds the optimal number of clusters, by performing model selection.

Correlation clustering is particularly appealing for the task of clustering structured objects, where the similarity function is domain-specific and can be designed or learned. A typical application is clustering web pages based on similarity scores. Moreover, correlation clustering is applicable to a multitude of other problems in different domains, including duplicate detection and similarity joins [Demaine et al., 2006, Hassanzadeh et al., 2009, Kushagra et al., 2019], spam detection [Bonchi et al., 2014a, Ramachandran et al., 2007], co-reference resolution [McCallum and Wellner, 2005], biology [Ben-Dor et al., 1999], image segmentation [Kim et al., 2011], social networks [Bonchi et al., 2015], clustering aggregation [Gionis et al., 2007], and experiment design [Pouget-Abadie et al., 2019]. Section 1.9 describes some of these.

1.2   EARLY WORK ON CORRELATION CLUSTERING

Correlation clustering has been rediscovered multiple times. The earliest formulation we are aware of dates back to 1964, when Zahn [1964] proposed the problem of finding an equivalence relation which "best approximates" a given symmetric relation (which is equivalent to Problem 1.1), and solved it for a special class of graph representing two-level and three-level hierarchies. Shortly thereafter, Régnier [1965] proposed a mathematical program for searching an equivalence relation that best approximates several given equivalence relations. This problem can be cast as a CLUSTERING-AGGREGATION problem, and hence reduces to correlation clustering (Section 1.8). Such "median" equivalence relations were extensively studied later [Mirkin, 1974, 1979], [Ambrosi, 1984, Barthelemy and Monjardet, 1981, Opitz and Schader, 1984], [Marcotorchino and Michaud, 1981a,b].

The first NP-hardness results for correlation clustering in this language were given by Křivánek and Morávek [1986] and Wakabayashi [1986]. In the context of clique partitioning (see Section 1.8), Grötschel and Wakabayashi [1989] gave an exact cutting-plane algorithm based on a natural integer programming formulation, and Grötschel and Wakabayashi [1990] studied facet-defining inequalities of the clique partitioning polytope. Later, motivated by problems in biological data analysis, Ben-Dor et al. [1999] considered the setting in which a graph is generated stochastically from a cluster graph H of genes by flipping the sign of each edge independently with a certain probability, and gave an efficient algorithm to obtain a clustering whose cost is at most that of H. Cohen and Richman [2001, 2002] studied the entity-resolution problem, in which it is required to put together the entries that correspond to the same entity. They suggested learning a pairwise similarity metric from past document data, and then partitioning a new set of documents in a way that correlates with it as much as possible. They proposed a simple randomized greedy algorithm which became the basis for many later developments.

The systematic study of the computational complexity of correlation clustering was initiated by Bansal et al. [2004] (who coined the term) and Shamir et al. [2004]. Both works independently showed NP-hardness of the binary case (Problem 1.1) and provided the first approximation algorithms. Since then, a plethora of other works extending and/or improving the basic formulation and exploring its practical applications have appeared.

1.3   COMPUTATIONAL HARDNESS

Problem 1.1 represents the simplest variant of correlation clustering. Even in this setting, finding an exact solution is computationally intractable since Problem 1.1 is NP-hard. This was shown by Bansal et al. [2004], Shamir et al. [2004], and previously, in related contexts, by Křivánek and Morávek [1986] and Wakabayashi [1986].

To show this, consider the decision version of Problem 1.1, which takes as input V, the similarity function s, and a parameter $K \in \mathbb{N}$, and asks if there exists a clustering $\ell : V \to \mathbb{N}$ with cost at most K.

Theorem 1.2 [**Bansal et al., 2004, Křivánek and Morávek, 1986, Shamir et al., 2004, Wakabayashi, 1986**]. *The decision problem of correlation clustering with binary similarities, i.e., in complete unweighted graphs (Problem 1.1), is* **NP**-*complete.*

Proof. That the problem is in **NP** is straightforward since, given a clustering ℓ, we can compute in polynomial time the number of disagreements and verify if it does not exceed K. To show **NP**-hardness, it suffices to give a polynomial-time reduction from another **NP**-hard problem. We follow the arguments from Bansal et al. [2004]. The reduction is from the PARTITION-INTO-TRIANGLES problem.

Problem 1.3 Partition-into-Triangles   Given a graph G with $n = 3k$ vertices, does there exist a partition of the vertices into K sets $V_1, \ldots, V_K$ such that for all i, $|V_i| = 3$ and the vertices in V_i form a triangle in G?

This problem is known to be **NP**-hard [Garey and Johnson, 1979]. Given an instance (G, K) of PARTITION-INTO-TRIANGLES, we need to construct in polynomial time an instance H of CORRELATION-CLUSTERING such that H has a clustering of cost $\leq K$ if and only if G has a partition into triangles.

First, observe that if G admits a partition into triangles, then the clustering corresponding to this triangulation has no negative mistakes (but it may have positive mistakes due to edges between triangles). Furthermore, any other clustering with clusters of size at most 3 has more positive mistakes than this clustering. We will augment the graph G to a larger complete graph H and design a gadget that forces the optimal correlation clustering of H to contain at most 3 vertices in each cluster.

The construction of H is as follows: H contains all the vertices and edges of G (the latter are labeled "+" in H). In addition, for every 3-tuple $\{u, v, w\} \subseteq G$, H contains a clique $C_{u,v,w}$ containing n^6 vertices. All edges inside these cliques are labeled "+". Furthermore, for all $u, v, w \in G$ each vertex in $C_{u,v,w}$ has a positive edge to u, v, and w, and a negative edge to all vertices in H not in $C_{u,v,w} \cup \{u, v, w\}$.

Now let us examine the structure of an optimal correlation clustering of H.

Lemma 1.4 [Bansal et al., 2004]   *The optimum correlation clustering of H has total cost at most* $K = n^7(\binom{n-1}{2} - 1) + \binom{n}{2}$.

Proof.   Consider a clustering C of H of the following form:

- There are $\binom{n}{3}$ clusters.

- Each cluster contains exactly one clique $C_{u,v,w}$ and possibly some vertices of G.

In any such clustering, there are no mistakes among edges between cliques (as G admits a triangulation). The only mistakes are between vertices of G and the cliques, and those between the vertices of G. The number of mistakes among vertices of G is at most $\binom{n}{2}$, and so the total number of mistakes of C is at most $K = n \cdot n^6(\binom{n-1}{2} - 1) + \binom{n}{2}$ because each vertex in G has n^6 positive edges to $\binom{n-1}{2}$ cliques and is clustered with only one of them.   □

Lemma 1.5 [Bansal et al., 2004]   *Any clustering of H having at least four vertices of G in some cluster has cost at least* $K' = n^7(\binom{n-1}{2} - 1) + \frac{n^6}{2} > K$.

Proof.   Consider a clustering X in which some cluster has four vertices u, v, w, y in G. First, without loss of generality, we can assume that each cluster in X has size at most $n^6 + n^4$, otherwise there are at least $\Omega(n^{10}) > K'$ negative mistakes within a cluster (because at most n^6 of those belong to the same clique $C_{u,v,w}$). This implies that each vertex in G makes at least $\binom{n-1}{2}n^6 - (n^6 + n^4)$ positive mistakes. Hence, the total number of positive mistakes is at least $n^7(\binom{n-1}{2} - 1) - n^5$. Let X_u be the cluster containing $u, v, w, y \in G$. Since X_u has at most $n^6 + n^4$ vertices, at least one of u, v, w, y will have at most n^4 positive edges inside X_u and hence will contribute at least an additional $n^6 - n^4$ negative mistakes to the clustering. Thus, the total number of mistakes is at least $(\binom{n-1}{2} - 1)n^7 - n^5 + n^6 - n^4 \geq n^7(\binom{n-1}{2} - 1) + \frac{n^6}{2}$. $\qquad\square$

Lemma 1.5 implies that the optimum correlation-clustering cost is at most K. Lemmas 1.4 and 1.5 together imply that G has a partition into triangles if and only if the cost of the optimum correlation clustering of H is at most K, thus completing the proof of Theorem 1.2.
$\square$

1.4   A MORE GENERAL FORMULATION

In many practical applications, the similarity information is not binary-valued, or it may be incomplete (i.e., for some pairs u, v, no cost has to be paid for putting u and v together or into different clusters). The framework from Section 1.1 naturally extends to these settings.

In more detail, the one by Ailon et al. [2008a] is, to date, the most general weighted formulation of correlation clustering. It assumes that each edge e is assigned two nonnegative weights, w_e^+ and w_e^-, and a clustering incurs cost w_e^+ if e is placed between clusters, and incurs cost w_e^- if e is placed within a cluster. If no restrictions are imposed on the weights w_e^+ and w_e^-, then it may happen that $w_e^+ = w_e^- = 0$. This is clearly equivalent to model the case that edge e is absent, meaning that the input graph is a general one. Every (w_{uv}^+, w_{uv}^-) pair expresses the advantage of putting u and v in the same cluster (w_{uv}^+) or in separate clusters (w_{uv}^-). The objective is to partition V so as to minimize the number of (weighted) disagreements, i.e., sum of the negative weights between objects within the same cluster plus the sum of the positive weights between objects in separate clusters.

Problem 1.6   Min-disagreement Correlation Clustering (Min-Disagree)
Given a set V of objects, and a pair $(w_{uv}^+, w_{uv}^-) \in \mathbb{R}^{\geq 0} \times \mathbb{R}^{\geq 0}$ of nonnegative weights for each $u, v \in V$, find a function $\ell : V \to \mathbb{N}$ so as to minimize

$$\text{cost}(l) = \sum_{\substack{(u,v)\in\binom{V}{2} \\ \ell(u)=\ell(v)}} w_{uv}^- + \sum_{\substack{(u,v)\in\binom{V}{2} \\ \ell(u)\neq\ell(v)}} w_{uv}^+. \qquad (1.2)$$

Assumptions.   We have assumed *nonnegative weights*. This makes sense since shifting each weight by an additive constant to each weight does not change the optimal solution. Note also that standard notions of multiplicative approximations would cease to make sense if negative weights were allowed.

If no additional condition is imposed on the weights, one may also assume that for any pair $u, v \in V$, at most one of w_{uv}^+ and w_{uv}^- is non-zero. There is no loss of generality in this because otherwise we may replace each weight pair w_{uv}^+ and w_{uv}^- by $w_{uv}'^+ = w_{uv}^+ - t_{uv}$ and $w_{uv}'^- = w_{uv}^- - t_{uv}$, where $t_{uv} = \min(w_{uv}^+, w_{uv}^-)$. Then Equation (1.2) becomes $\sum_{(u,v)} 2t_{uv} + \sum_{\ell(u)=\ell(v)} w_{uv}^- + \sum_{\ell(u) \neq \ell(v)} w_{uv}^+$. As the first term is constant for any given input and all terms are nonnegative, any α-approximate minimizer of this expression gives an α-approximate solution to Problem 1.6. Thus, we may think of the input weights as given by an *undirected*, weighted, simple, and complete graph $G = (V, E)$ whose edges are labeled either $+$ or $-$ (but not both).

However, in many problems, a number of additional conditions (restrictions) on the weights may apply. This fact can often lead to improved approximation algorithms. The most important ones are listed below.

Complete vs. incomplete graphs.   When $w_{uv}^+ + w_{uv}^- > 0$ for all u, v, we say that we are performing correlation clustering on *complete graphs* or with complete similarity information. If instead $w_{uv}^+ = w_{uv}^- = 0$ holds for some pairs, we say we are dealing with *incomplete graphs*.

Probability constraints.   A special case of complete similarity occurs when the weights satisfy the *probability constraint*: $w_{uv}^+ + w_{uv}^- = 1$ for all u, v. Intuitively, the probability constraints ensure that the same "importance" is given to each pair of vertices in the graph. In this case, a single parameter suffices to describe an edge (e.g, $x_{uv} = \frac{1}{2}(w_{uv}^+ - w_{uv}^-) \in [-1, 1]$). If, in addition, the weights take values in $\{0, 1\}$ we recover Problem 1.1 (where $s(x, y) = w_{xy}^+$).

Triangle-inequality constraints.   These are satisfied when $w_{uv}^- + w_{vw}^- \geq w_{uw}^-$. This is the case whenever the negative weights are given by a metric; typical examples include Euclidean distance, one minus Jaccard similarity, and the square root of Pearson dissimilarity.

From binary weights to probability constraints.   It turns out that the case of weights with probability constraints can be reduced to the binary case, at the cost of a small factor in the approximation.

Theorem 1.7 [Bansal et al., 2004].   *Any k-approximation algorithm for correlation-clustering in complete graphs with binary weights yields a $(2k + 1)$-approximation algorithm to the weighted problem under probability constraints.*

Proof.   Let G be a graph encoding the weighted correlation clustering instance with probability constraints and let G' be the unweighted graph obtained from G by the following "thresholding" procedure: if $w_e^+ > \frac{1}{2}$ we put a positive edge between the endpoints of e; if $w_e^+ < \frac{1}{2}$ we put a negative edge between the endpoints of e (edges with $w_e^+ = \frac{1}{2}$ are signed arbitrarily). Let OPT be the optimal clustering of G and OPT' the optimal clustering of G'. Also, let cost and cost' denote the cost functions in G and G'.

Given any clustering C, consider an edge e on which C incurs a greater penalty in cost' than in cost. The penalty in cost' must be 1, hence e is either a positive edge between clusters with $w_e^+ \geq \frac{1}{2}$, or a negative edge inside a cluster with $w_e^- \leq \frac{1}{2}$. In either case, cost incurs a cost of at least $\frac{1}{2}$. Thus, $\text{cost}'(\text{OPT}') \leq \text{cost}'(\text{OPT}) \leq 2\,\text{cost}(\text{OPT})$, where the first inequality holds by design (definition of OPT').

Suppose we have an algorithm $\mathcal{A}'$ to find k-approximate solutions to correlation clustering in complete binary graphs. Our algorithm $\mathcal{A}$ simply runs $\mathcal{A}'$ on G' and outputs the resulting clustering A. Then $\text{cost}'(A) \leq k\,\text{cost}'(\text{OPT}') \leq 2k\,\text{cost}(\text{OPT})$.

Now consider $\text{cost}(A) - \text{cost}'(A)$. Any positive edge between clusters, or negative edge inside a cluster, contributes at most 0 to this difference. All other edges contribute at most $\text{cost}(\text{OPT})$. Therefore, $\text{cost}(A) \leq \text{cost}'(A) + \text{cost}(\text{OPT})$, which yields $\text{cost}(A) \leq \text{cost}'(A) + \text{cost}(\text{OPT}) \leq (2k + 1)\,\text{cost}(\text{OPT})$. $\qquad\qquad\square$

1.4.1 MAX-AGREEMENT CORRELATION CLUSTERING

An alternative formulation attempts to maximize the sum of the positive weights between objects within the same cluster plus the sum of the negative weights between objects in separate clusters (MAX-AGREEMENT CORRELATION CLUSTERING, MAX-AGREE for short).

Problem 1.8 Max-agreement Correlation Clustering (Max-Agree)
Given a set V of objects, and a pair $(w_{uv}^+, w_{uv}^-) \in \mathbb{R}_0^+ \times \mathbb{R}_0^+$ of nonnegative weights vectors where $w_{uv} = w_{vu} \; \forall u, v \in V$, find a function $\mathcal{C} : V \to \mathbb{N}^+$ so as to maximize

$$\sum_{\substack{u,v \in V \\ \mathcal{C}(u) = \mathcal{C}(v)}} w_{uv}^+ \; + \sum_{\substack{u,v \in V \\ \mathcal{C}(u) \neq \mathcal{C}(v)}} w_{uv}^-. \qquad (1.3)$$

Both problems MIN-DISAGREE and MAX-AGREE are equivalent from the point of view of exact optimization because the sum of their objective values is a constant independent of the clustering ℓ. However, it is conceptually helpful to distinguish between them because they differ from an approximation viewpoint. Roughly speaking, good multiplicative approximations to MIN-DISAGREE require finding near-optimal solutions when the optimal number of disagreements is very low, whereas good multiplicative approximations to MAX-AGREE require finding near-optimal solutions when the optimal number of disagreements is very high. In practical applications, it is MIN-DISAGREE that tends to be more important because clustering algorithms are usually applied only when some semblance of a low-cost clustering appears possible.

By way of example, consider the complete setting with binary weights. Then a 2-factor approximation algorithm for MAX-AGREE is trivial: a random clustering into two clusters will agree with at least half the edges. On the other hand, it is unknown whether MIN-DISAGREE admits a polynomial-time 2-factor approximation. At the other extreme, when the optimal solution to the input instance has cost, say, $\geq n^2/3$, then any clustering gives a 3/2-factor approximation to the minimum number of disagreements, but not necessarily a 3/2-factor approximation to the maximum number of agreements.

1.5   APPROXIMATION ALGORITHMS FOR MIN-DISAGREE

1.5.1   APPROXIMATION ALGORITHMS FOR COMPLETE GRAPHS

The first algorithm with provable guarantees for MIN-DISAGREE on complete graphs was a 17,429-approximation given by Bansal et al. [2004]. Later, Charikar et al. [2005] devised a 4-approximation algorithm for the same problem, based on rounding the solution of a linear program. Here we describe instead the simpler (randomized) QwickCluster algorithm of Ailon et al. [2008a], which also achieves a better approximation (but not the best one known, as we shall see). The idea behind QwickCluster is that in a MIN-DISAGREE instance with a good (low-cost) solution, the majority of the vertices in the graph are involved in a small number of disagreements (because there are few disagreements overall). Therefore, if we pick a random vertex from the graph, we should be able to recover most of the cluster it belongs to in the optimal clustering (plus some noise) simply by looking at its neighbors. Following this intuition, the QwickCluster algorithm selects a random cluster center or "pivot," creates a cluster with it and its positive neighborhood, removes the cluster, and iterates on the remaining graph, until we are left with an empty graph. Essentially, it finds a maximal independent set in the graph of positive edges in random order, which serves as a set of pivots.

Algorithm 1.1 QwickCluster [Ailon et al., 2008a]

Input: $G = (V, E)$, a complete graph with "+,−" edge labels
  $R \leftarrow V$ // Unclustered vertices so far
  while $R \neq \emptyset$ **do**
    Pick a pivot v from R uniformly at random.
    Output cluster $C = \{v\} \cup \Gamma_G^+(v) \cap R$.
    $R \leftarrow R \setminus C$
  end while

Theorem 1.9 [Ailon et al., 2008a]. *The expected cost of QwickCluster is at most three times the cost of the optimum clustering.*

In order to analyze Algorithm 1.1, we need to compare the expected cost of the clustering it outputs with the cost of the (unknown) optimum solution OPT. We do this by identifying a quantity L that serves as a good lower bound on the cost of OPT, and showing that the expected cost of Algorithm 1.1 is at most $3L$. That is, we will show that

$$L \leq \text{cost(OPT)} \leq 3 \, \mathbb{E}[\text{cost(QwickCluster)}].$$

The key concept is that of a *bad triangle*: a triangle in G with two positive edges and a negative edge in the input graph. Notice that every bad triangle forces the existence of at least one disagreement involving one of its three edges, for any clustering. Let $B \subseteq \binom{V}{3}$ denote the set of all bad triangles.

Observation 1: *The set of disagreements caused by any clustering must intersect every bad triangle.* Hence the cost of any clustering is at least as large as the *smallest hitting set* for B, i.e., the smallest set of edges intersecting every bad triangle. Thus, we can define a valid lower bound $\hat{L}$ for the optimum cost by

$$\hat{L} \triangleq \min\{|R| \mid R \cap T \neq 0 \; \forall T \in B\} \leq \text{cost(OPT)}. \tag{1.4}$$

The inequality in (1.4) may be strict, because there is no guarantee that the smallest hitting set for B can actually be realized as the set of disagreements of some clustering.

The following integer program formulates the problem of computing $\hat{L}$:

$$
\begin{aligned}
\hat{L} = \min \quad & \sum_{e \in E} x_e \\
\text{s.t.} \quad & \sum_{e \in E} x_{e_1} + x_{e_2} + x_{e_3} \geq 1 \qquad \forall \text{ bad triangle } \{e_1, e_2, e_3\} \in B \\
& x_e \in \{0, 1\} \qquad \forall e \in E.
\end{aligned} \tag{1.5}
$$

Observation 2: *We can obtain a weaker (smaller) lower bound by relaxing the binary constraints $x_e \in \{0, 1\}$ in (1.5).* We simply require instead that each x_e be a nonnegative real; the advantage of doing so is that now we obtain a linear program, which is easier to analyze than the integer program (1.4).

$$
\begin{aligned}
\min \quad & \sum_{e \in E} x_e \\
\text{s.t.} \quad & \sum_{e \in E} x_{e_1} + x_{e_2} + x_{e_3} \geq 1 \quad \forall \text{ bad triangle } \{e_1, e_2, e_3\} \in B \\
& x_e \geq 0 \qquad \forall e \in E.
\end{aligned} \tag{1.6}
$$

Notice that there would be no benefit to adding the upper-bound constraints of the form $x_e \leq 1$, because they would be automatically satisfied in any optimal solution.

The optimum value of (1.6) represents the smallest fractional covering of all bad triangles by fractional edges. By linear-programming duality, it equals the optimum value of the dual linear program, which represents the largest fractional packing of bad triangles such that no edge appears in more than one triangle in total:

$$
\begin{aligned}
L = \max \quad & \sum_{y \in T} y_T \\
\text{s.t.} \quad & \sum_{T \in B \mid e \in T} y_T \ \leq \ 1 \quad \forall \text{ edge } e \in E \\
& \qquad\qquad y_T \ \geq \ 0 \qquad \forall T \in B.
\end{aligned}
\tag{1.7}
$$

Any feasible solution to (1.7) yields a lower bound for L and hence for $\mathrm{cost}(\mathsf{OPT})$:

$$
L \leq \hat{L} \leq \mathrm{cost}(\mathsf{OPT}).
\tag{1.8}
$$

Observation 3. *Any edge disagreement in the output of the algorithm can be attributed to exactly one bad triangle.* Indeed, if the edge $e = (i, j)$ is positive but causes an error, it must be because i and j are clustered separately, which happens precisely when, at some point during the execution of the algorithm, both i and j are unassigned but a third vertex k which is a neighbor of exactly one of the two is chosen as pivot. On the other hand, if the edge is negative but causes an error, it must be because i and j are clustered together, which happens precisely when, at some point during the execution of the algorithm, both i and j are unassigned but a third vertex k which is a neighbor of the two is chosen as pivot. Whenever one of these triangle events happens, we incur an error only in one of the three edges of the triangle (i, j, k).

For each bad triangle $t \in B$, let A_t be the event that the algorithm chooses one of the three vertices in t as a pivot while the other two are still unassigned. It follows from the remark above that the number of disagreements is $|\{t \in B \mid A_t \text{ occurs}\}| = \sum_{t \in B} \mathbb{1}[A_t]$, so the expected cost of the clustering returned by QwickCluster is given by

$$
\mathbb{E}[\mathrm{cost}(\mathsf{QwickCluster})] = \sum_{t} \Pr[A_t].
\tag{1.9}
$$

Observation 4. *We can obtain a high-value feasible solution to LP* (1.7). To this end, let $p_t = \Pr[A_t]$; then the assignment $y_t = p_t/3 \geq 0$ is feasible for LP (1.7). To see this, observe that QwickCluster selects a pivot uniformly at random among unassigned vertices, so conditioned on the event A_t, each of the three vertices in T is equally likely to be selected as pivot. Let B_e denote the event that $e \in T$ becomes an error in the clustering. Then the probability that both B_e and A_t happen is $\frac{p_t}{3}$. But each bad edge can only be charged to one triangle incident to it, so the probability that e becomes wrong is precisely $\sum_{t \mid e \in t} \frac{p_t}{3}$. Hence, the last sum is at most 1 as required by LP (1.7).

It follows that our lower bound L is bounded by the value of the solution $\{p_t\}$:

$$L \geq \frac{1}{3} \sum_{t \in B} \Pr[A_t]. \tag{1.10}$$

Combining (1.8), (1.9), and (1.10), we see that

$$\text{cost(OPT)} \geq L \geq \frac{1}{3} \sum_{t \in B} \Pr[A_t] = \frac{1}{3} \, \mathbb{E}[\text{cost(QwickCluster)}],$$

which completes the proof. □

1.5.2   EXTENSION TO PROBABILITY CONSTRAINTS

By Theorem 1.7, we know that the 3-approximation given by QwickCluster can be used to obtain a 7-approximation algorithm for graphs satisfying the probability constraints by applying a thresholding procedure to the graph. In fact, Ailon et al. [2008a] analyzed the behavior of QwickCluster in the resulting binary graph and showed that it gives a 5-approximation to the original problem in this case. If, in addition to the probability constraints, the triangle inequality constraints also hold, then it gives a 2-approximation.

While these algorithms are randomized, van Zuylen and Williamson [2007, 2009] showed how to match the same approximation guarantees with a deterministic algorithm. The pivot selection step is derandomized by choosing a pivot so as to minimize a ratio of expectations; then they apply the method of conditional expectations to decide which other elements to cluster together with the pivot.

A generalization of the probability constraint is taken into account by Puleo and Milenkovic [2015]. See Section 2.2 for more details on this.

A further generalization is considered by Mandaglio et al. [2021], who initiate the study of correlation clustering with *global weight bounds*, i.e., constraints to be satisfied by the input weights altogether, as opposed to the *local* constraints (such as, e.g., probability constraints or Puleo and Milenkovic [2015]'s constraints) that are required to hold for every object pair.

Improving the approximation via LP rounding.   The results above are not the best possible. To improve on them, consider the following natural formulation of MIN-DISAGREE with general weights. Let us x_{uv} denote a binary "distance" between objects, so that $x_{uv} = 0$ if u and v should be clustered together and $x_{uv} = 1$. By symmetry, $x_{uv} = x_{vu}$. By transitivity, if $x_{uv} = 0$ and $x_{vw} = 0$, then we must have $x_{uw} = 0$; this condition may be written as $x_{uv} + x_{vw} \geq x_{uw}$ if we allow linear constraints. It is easy to see that any such solution with $x_{uv} \in \{0, 1\}$ represents a clustering with cost $\sum_{u,v} w_{uv}^-(1 - x_{uv}) + w_{uw}^+ x_{uw}$, which should be our objective function to minimize.

Now suppose that, in order to make the program tractable, we relax the condition $x_{uv} \in \{0, 1\}$ to $x_{uv} \in [0, 1]$. Then we arrive at the following linear program which lower bounds the cost of any optimal clustering:

$$\min \quad \sum_{u,v} w_{uv}^{-}(1 - x_{uv}) + w_{uw}^{+} x_{uw}$$

$$\text{s.t.} \qquad\qquad x_{uv} + x_{vw} \geq x_{uw}, \qquad\qquad (1.11)$$
$$x_{uv} = x_{vu},$$
$$x_{uv} \in [0, 1].$$

This program can be solved in polynomial time using an LP solver. Note that the transitivity condition looks like a triangle inequality $x_{uv} + x_{vw} \geq x_{uw}$, thus any solution to LP (1.11) can be interpreted as encoding a semi-metric between the clusters. What remains to be seen is how to *round* the distances x_{uv} into binary values without incurring a too high increase in the objective function.

The key insight by Ailon et al. [2008a] is that we can use QwickCluster to this end. The only modification required is that, when we pick a pivot u at random, rather than putting it together with all its positive neighbors, we will put it together with each remaining unclustered vertex v with probability $1 - x_{uv}$.

Algorithm 1.2 LP-QwickCluster [Ailon et al., 2008a]

Input: V equipped with a distance function x_{uv}
   $R \leftarrow V$ // Unclustered vertices so far
  while $R \neq \emptyset$ **do**
     Pick a pivot v from R uniformly at random.
     $C \leftarrow \{v\}$
     for all $w \in R \setminus \{v\}$ **do**
       With probability $1 - x_{vw}$, set $C = C \cup \{w\}$.
     end for
     Output cluster C.
     $R \leftarrow R \setminus C$.
  end while

Theorem 1.10 [Ailon et al., 2008a].   *Suppose the weights w obey the probability constraints. If x denotes the optimal solution to LP (1.11), LP-QwickCluster returns a 2.5-approximation to the optimal correlation clustering solution.*

Again, the same guarantee holds for a derandomized version of the algorithm using the techniques of van Zuylen and Williamson [2009].

The best approximation factor known for MIN-DISAGREE with binary complete graphs was obtained by Chawla et al. [2015]. Their approach adds a third, intermediate step to the above scheme from Ailon et al. [2008a]. Rather than using the x_{uv} distances computed in LP (1.11) directly as the probability of separating a pivot u from a vertex v, they introduce two rounding functions, f^+ and f^-, mapping distances x_{uv} to separation probabilities; the function to be used depends on whether (u, v) form a positive or a negative edge. These functions are degree-2 polynomials in x_{uv} whose constants are to be optimized in order to achieve the best approximation ratio. Their approach extends also to weighted graphs as long as the triangle inequality and probability constraints hold. Their main result is summarized below.

Theorem 1.11 [Chawla et al., 2015]. *There is a polynomial-time 2.06-approximation algorithm for* Min-Disagree *in binary complete graphs, and a 1.5-approximation algorithm for* Min-Disagree *in complete graphs with edge weights satisfying triangle inequalities and probability constraints.*

It is interesting to note that these approximation ratios are close to the best ratio possible using techniques based on rounding LP (1.11), since the integrality gap of (1.11) with binary weights is known to be at least 2 [Charikar et al., 2005], and, if the weights satisfy triangle inequalities and probability constraints, the integrality gap is at least 1.2 [Chawla et al., 2015].

1.5.3   AN $O(\log n)$-APPROXIMATION FOR GENERAL WEIGHTED GRAPHS

For incomplete graphs and general weighted graphs, the situation is different and the problem becomes harder. Several authors have noticed that Min-Disagree in this setting is equivalent to the (weighted) Min-Multicut problem [Charikar et al., 2005, Demaine et al., 2006]. This equivalence yields both an $\mathcal{O}(\log n)$-approximation algorithm and a hardness result for Min-Disagree.

The Min-Multicut problem is the following: given an undirected graph G, a weight function w on the edges of G, and a collection of k pairs of distinct vertices (s_i, t_i) of G, find a minimum-weight set of edges of G whose removal disconnects every s_i from the corresponding t_i.

From correlation clustering to weighted multicut.   First, we give a reduction from the Min-Disagree problem Min-Multicut. The starting point is to generalize the notion of bad triangle to that of *bad cycles*: a simple cycle containing exactly one negative edge. It is easy to show that the optimum Min-Disagree cost is equal to the smallest weighted cost of edges needed to remove all bad cycles. This generalizes Observation 1 from Section 1.5, because, for complete graphs, the existence of a bad cycle implies the existence of a bad triangle. Given a weighted graph G whose edges are labeled "+" or "−", we construct a new graph H_G by keeping all vertices and all positive edges, and replacing every negative edge $e = (u, v)$ by the following gadget. Let us add a new vertex z_e, a new edge (z_e, u) with weight equal to that of e, and a source-sink pair (z_e, v). Let S_G denote the set of source-sink pairs.

Lemma 1.12 [**Demaine et al., 2006**] (H_G, S_G) *has a cut of weight* W *if and only if* G *has a clustering of weight* W, *and we can easily construct one from the other.*

The reduction from Lemma 1.12 preserves the value of the objective function. More specifically, we have presented a mapping f from an instance x of the MIN-DISAGREE problem to an instance $f(x)$ of the MIN-MULTICUT problem, and a mapping g from a feasible solution of $f(x)$ (MIN-MULTICUT problem) to a feasible solution of x (MIN-DISAGREE problem) such that the value of the solution does not change. This implies that approximate solutions to MIN-MULTICUT translate into approximate solutions to MIN-DISAGREE as well. The MIN-MULTICUT problem with k source-sink pairs has an $\mathcal{O}(\log k)$ approximation [Garg et al., 1996]. Hence, we obtain the following.

Theorem 1.13 [**Demaine et al., 2006**].   MIN-DISAGREE *in general weighted graphs has a* $\mathcal{O}(\log n)$ *polynomial-time approximation algorithm.*

From weighted multicut to correlation clustering.   Consider an input to the MIN-MULTICUT problem, given by an undirected graph H, a weight function $w : E \to \mathbb{R}^+$ on the edges of H, and a collection of k pairs of distinct vertices $S = \{(s_1, t_1), \ldots, (s_k, t_k)\}$. We construct a MIN-DISAGREE instance in polynomial time as follows:

- We start with $G_H = H$, all edge weights are preserved and all edges labeled "+".

- In addition, for every source-sink pair (s_i, t_i), we add to G_H a negative edge $e_i = (s_i, t_i)$ with weight $w(e_i) = \sum_{e \in H} w(e) + 1$.

The following result is easily shown.

Lemma 1.14 [**Demaine et al., 2006**]   *A clustering on* G_H *with weight* W *induces a multicut on* (H, S) *with weight at most* W. *Similarly, a multicut in* (H, S) *with weight* W *induces a clustering on* G_H *with weight* W.

As a consequence we obtain.

Theorem 1.15 [**Demaine et al., 2006**].   *The* MIN-DISAGREE *problem is APX-hard, even in the unweighted case.*

Finally, such an equivalence between the MIN-MULTICUT problem and MIN-DISAGREE on general graphs also implies that the $\mathcal{O}(\log n)$ approximation for MIN-DISAGREE is unlikely to be improvable. In fact, any improvement on the approximation of MIN-DISAGREE would also result in an improvement on the approximation of MIN-MULTICUT. This is unlikely since, assuming the Unique Games Conjecture [Khot, 2002], no constant-factor approximation algorithm may exist for MIN-MULTICUT [Chawla et al., 2006].

1.6 APPROXIMATION ALGORITHMS FOR MAX-AGREE

MAX-AGREE is easier to approximate than MIN-DISAGREE. By way of example, consider binary graphs. If we partition the vertices into two clusters at random, then every edge causes an agreement with probability half, therefore the total expected number of agreements is equal to half the number of edges. Hence, a random splitting into two clusters yields a trivial 0.5-approximation in expectation. A similar argument actually applies to weighted graphs. Moreover, the method can be derandomized: consider the clustering C_1 that puts all vertices into one big cluster and the clustering C_n which puts all vertices into separate clusters. Let W be the total weight of all edges. If at least half the total weight is in positive edges, then C_1 has at least $W/2$ weighted agreements; while if at least half the edge labels are negative, then C_n has at least $W/2$ weighted agreements. Thus, taking best of the two yields always a 0.5-approximation.

In fact, unlike MIN-DISAGREE, for any fixed $\epsilon > 0$ MAX-AGREE can be approximated in complete binary graphs to within a $1 + \epsilon$ factor in polynomial time; that is, there exists a polynomial-time approximation scheme (PTAS) for it. This was shown by Bansal et al. [2002], using techniques developed by de la Vega and Kenyon [1998] to obtain a PTAS for MAX-CUT. The running time is $\mathcal{O}(n^2 \exp(O(1)/\epsilon))$. The results from Giotis and Guruswami [2006a], presented in Section 2.1, also imply a randomized PTAS for MAX-AGREE with running time $\mathcal{O}(n \cdot (\frac{1}{\epsilon})^{\epsilon^{-3} \log(k/\epsilon)})$. A faster PTAS for MAX-AGREE was obtained by Bonchi et al. [2013a], using a local algorithm as outlined in Section 5.2; it runs in time $\mathcal{O}(n \cdot \epsilon^{-\mathcal{O}(1)} + \exp(\mathcal{O}(1/\epsilon)))$.

For general weighted graphs, the problem becomes APX-hard by reduction from MAX-3SAT [Charikar et al., 2005], thus a PTAS is ruled out unless $P = NP$. Recall that the randomized 0.5-approximation described in the beginning of this section opens two clusters and achieves a 0.5-factor approximation. This can be improved upon, and interestingly, it turns out that opening a small number of clusters is always near-optimal for MAX-AGREE. Specifically, both Charikar et al. [2005] and Swamy [2004] independently proposed a natural semidefinite-programming (SDP) formulation of MAX-AGREE inspired by the breakthrough approximation algorithm for MAX-CUT of Goemans and Williamson [1995]. Charikar et al. [2005] presented a rounding scheme for such a semidefinite program which achieves a 0.7664-approximation by opening at most 8 clusters, whereas Swamy [2004] showed a rounding scheme for such an SDP which achieves a 0.75-approximation by opening at most 4 clusters, and another one which achieves a 0.7666-approximation by opening at most 6 clusters. The latter remains to date the best approximation factor known for MAX-AGREE.

1.7 CORRELATION CLUSTERING IN PRESENCE OF A GROUND TRUTH

Correlation clustering has traditionally been considered from an *agnostic* perspective, in the sense that the existence of a correct clustering is not taken into account: the cost of a solution is computed only against the input edge-labeling function $\mathcal{L}$, which states whether a certain pair

(u, v) of vertices should ideally lie within the same cluster $(\mathcal{L}(u, v) = +1)$ or not $(\mathcal{L}(u, v) = -1)$. Ailon and Liberty [2009] fill this gap and study a variant of correlation clustering where an unknown ground-truth clustering ℓ^* is assumed to exist, and the input edge-labeling function $\mathcal{L}$ is just a proxy of ℓ^*. This scenario arises, for instance, in duplicate detection and elimination in large datasets (also known as record linkage).

Specifically, the setting considered by Ailon and Liberty [2009] is an adversarial one: not only the ground-truth clustering ℓ^* is unknown, but it can also be arbitrarily different from the input edge-labeling function $\mathcal{L}$. Since an algorithm can only access $\mathcal{L}$, one can expect the output ℓ to well-represent the ground truth only insofar as the input $\mathcal{L}$ does. Therefore, the goal of Ailon and Liberty [2009] in this regard is to find a minimum constant W such that the number $\delta(\ell, \ell^*)$ of disagreements between ℓ and ℓ^* is no more than W times the number $\delta(\mathcal{L}, \ell^*)$ of disagreements between $\mathcal{L}$ and ℓ^*, i.e., $\delta(\ell, \ell^*) \leq W\delta(\mathcal{L}, \ell^*)$. In other words, the more $\mathcal{L}$ disagrees with the ground truth ℓ^* (larger $\delta(\mathcal{L}, \ell^*)$), the weaker the requirements from the output ℓ. Given that the ground-truth clustering ℓ^* is unknown, the problem ultimately studied by Ailon and Liberty [2009], which we hereinafter refer to as CORRELATION-CLUSTERING-WITH-GROUND-TRUTH, is aimed at finding a minimum constant W^* such that $\delta(\ell, \hat{\ell}) \leq W^*\delta(\mathcal{L}, \hat{\ell})$, *for all possible clusterings $\hat{\ell}$.*

Problem 1.16   (Correlation-Clustering-with-Ground-Truth)   Let $G = (V, E, \mathcal{L})$ be a signed graph, where V is a set of n vertices, $E \subseteq \binom{V}{2}$ is a set of m edges, and $\mathcal{L} : E \to \{-1, +1\}$ is a function labeling an edge as either positive or negative. Let also $\delta(\cdot, \cdot)$ denote the number of disagreements between two clusterings or between a clustering and an edge-labeling function. The objective is to find a constant $W^* \geq 1$ and a clustering $\ell : V \to \mathbb{N}$ so as to

$$\text{minimize}\quad W^*$$
$$\text{subject to}\quad \delta(\ell, \hat{\ell}) \leq W^*\delta(\mathcal{L}, \hat{\ell}), \text{for all clusterings } \hat{\ell}.$$

An important observation about Ailon and Liberty [2009]'s CORRELATION-CLUSTERING-WITH-GROUND-TRUTH problem is as follows. The traditional MIN-DISAGREE problem (where no ground-truth is taken into consideration) aims at finding a clustering ℓ that approximately minimizes $\delta(\ell, \mathcal{L})$. This means that constant-factor approximation algorithms for MIN-DISAGREE indirectly solve the problem of finding a constant $Z \geq 1$ such that $\delta(\ell, \mathcal{L}) \leq Z\delta(\hat{\ell}, \mathcal{L})$, for all possible clusterings $\hat{\ell}$. By triangle inequality, it holds that $\delta(\ell, \ell^*) \leq \delta(\hat{\ell}, \mathcal{L}) + \delta(\mathcal{L}, \ell) \leq (Z + 1)\delta(\mathcal{L}, \hat{\ell})$. Hence, an approximation factor of Z for MIN-DISAGREE gives an upper bound of $W^* = Z + 1$ for CORRELATION-CLUSTERING-WITH-GROUND-TRUTH. Since MIN-DISAGREE is APX-hard, then in the considered scenario $Z > 1$, and this approach would therefore yield a $W^* > 2$. A similar argument can be made for randomized combinatorial-optimization and the corresponding expected approximation ratios. This consideration raises the interesting question of whether one can go below $W^* = 2$ and shortcut the traditional optimization detour (often an obstruction under complexity theoretical assumptions).

The main result of Ailon and Liberty [2009] is a morphing process that proves the existence of a good relaxed solution to CORRELATION-CLUSTERING-WITH-GROUND-TRUTH, which implies, as a major consequence, the existence of a $W^* = \frac{4}{3} < 2$. Specifically, Ailon and Liberty [2009] show that one can continuously change the values of the input $\mathcal{L}$ into a "soft" clustering ℓ_{dif} which is [0, 1] valued and a metric (satisfying all triangle inequalities). The relaxation ℓ_{dif} is obtained as the limit at infinity of a solution to a piecewise linear differential equation. The algorithm devised for this task, which is also referred to as a morphing process, is interesting in its own right and may be useful for other problems on metric spaces. The intuitive idea behind the differential equation is a physical system in which edges "exert forces" on each other proportional to the size of triangle inequality violations. Within this view, it is shown that all triangle inequality violations decay exponentially in time, and this fast decay allows us to bound the loss with respect to the ground truth from above.

Another noteworthy result of Ailon and Liberty [2009] consists in showing how to randomly convert the good relaxed solution ℓ_{dif} into an integer solution ℓ to CORRELATION-CLUSTERING-WITH-GROUND-TRUTH such that $\delta(\ell, \hat{\ell}) \leq \frac{3}{2}\delta(\ell_{dif}, \hat{\ell})$. Applying the rounding algorithm to ℓ_{dif} gives a $W^* \leq 2$ approximation algorithm. As a side effect, Ailon and Liberty [2009]'s randomized algorithm allows for computing an invariant $W' = \leq \frac{4}{3}$, which serves as a witness for getting a solution to CORRELATION-CLUSTERING-WITH-GROUND-TRUTH with $W^* = \frac{3}{2}W'$. In particular, if $W' < \frac{4}{3}$, then $W^* < 2$.

The theory behind the morphing process(es) and the randomized algorithm devised by Ailon and Liberty [2009] is rather complex, we therefore omit it here. For details on that, we defer the interested reader to the original work by Ailon and Liberty [2009] and the extended technical report therein cited.

1.8 RELATED PROBLEMS

There exist several problems that, although not being framed as correlation clustering, are close to it. In the following we overview them.

Cluster-graph modification. The goal of *cluster-graph modification problems* is to transform an input graph into a *cluster graph*, i.e., a set of disjoint cliques, by performing a minimum number of edge modifications. Based on the type of modification permitted, three main formulations have been tackled in the literature: CLUSTER-COMPLETION, CLUSTER-DELETION, and CLUSTER-EDITING, where it is allowed to perform edge additions only, edge deletions only, or both, respectively [Shamir et al., 2004].

It is not difficult to see that CLUSTER-EDITING is equivalent to MIN-DISAGREE on complete graphs, while CLUSTER-COMPLETION and CLUSTER-DELETION are variants of MIN-DISAGREE where there cannot exist negative inter-cluster edges and a cost is paid for negative intra-cluster edges (CLUSTER-COMPLETION), or clusters can contain positive edges only and a cost is paid for positive inter-cluster edges (CLUSTER-DELETION). These problems are typically studied from the point of view of *parameterized complexity* of exact algorithms: given an upper bound k on the maximum number of modifications allowed, determine if there is a solution. It is possible to design exact algorithms for this that run in polynomial time for each *fixed k*. The running time of these algorithms will grow exponentially with k, but the base of the exponential dependence can be reduced via data reduction rules and kernelization techniques. For example, the fastest known strategy for cluster editing with k edges [Böcker et al., 2009] runs in time $\text{poly}(n) \cdot 1.82^k$. For all cluster-graph modification problems, several variants exist, including variants working on edge-weighted input graphs [Böcker et al., 2009], variants where the output solution is constrained to have a fixed number of clusters [Shamir et al., 2004], and variants handling other types of graphs, e.g., bipartite graphs [Amit, 2004].

For more detailed surveys on cluster-graph modification problems we defer to Böcker and Baumbach [2013] (focused on the CLUSTER-EDITING variant) and Crespelle et al. [2020] (focused on the broader class of edge-modification problem).

Edge-clique partition. The class of problem known as *edge-clique partition* or *clique clustering* consists in partitioning the vertices of an input graph into disjoint clusters such that each cluster forms a clique and the number of edges within the clusters is maximized (MAX-EDGE-CLIQUE-PARTITION variant), or the number of edges between clusters is minimized (MIN-EDGE-CLIQUE-PARTITION variant). The problems were introduced by Figueroa et al. [2005] in the context of DNA clone classification.

MIN-EDGE-CLIQUE-PARTITION is equivalent to the aforementioned CLUSTER-DELETION problem, while MAX-EDGE-CLIQUE-PARTITION is the maximization counterpart of it. Although the two variants are equivalent in terms of optimality (i.e., an exact solution to MAX-EDGE-CLIQUE-PARTITION is an exact solution to MIN-EDGE-CLIQUE-PARTITION, and vice versa), they possess different approximation properties [Dessmark et al., 2007, Punnen and Zhang, 2012, Sukegawa and Miyauchi, 2013]. Exact algorithms based on integer linear programming and SAT solvers have been proposed and shown to be feasible for small real-world datasets [Miyauchi et al., 2018]. The problems have also been studied in the online setting [Chrobak et al., 2020].

Clustering aggregation. Gionis et al. [2007] study the problem of CLUSTERING-AGGREGATION, that is, given a set of clusterings, find a single clustering (possibly not included in the given set) that agrees as much as possible with the input clusterings. More specifically, Gionis et al. [2007] consider as a measure of disagreement between any two clusterings the number of object pairs on which the two clusterings disagree, and formulate CLUSTERING-AGGREGATION as the problem of finding the clustering that minimizes the sum of the individual disagreement between it and all the input clusterings.

The connection between CLUSTERING-AGGREGATION and correlation clustering is that the former can be shown to be a restricted version of MIN-DISAGREE, with edge weights that are ad-hoc defined as the fraction of input clusterings where the endpoints of that edge have not been clustered together. The weights so defined are shown by Gionis et al. [2007] to satisfy the probability constraint and obey the triangle inequality. This allows Gionis et al. [2007] to devise algorithms for CLUSTERING-AGGREGATION by exploiting the well-established results of correlation clustering on probability-constraint- and triangle-inequality-compliant weighted graphs.

Rank aggregation.   The term *rank aggregation* refers to a class of problem whose broad goal is to combine an input set of complete ranked lists defined on the same set of elements into a single ranking, in such a way that the resulting ranking best describes the preferences expressed in the given lists. The problem dates back to as early as the late 18th century, when Condorcet [1785] and Borda [1781] each proposed voting systems for elections with more than two candidates. In the last half century, rank aggregation has been studied and defined from a mathematical perspective. In particular, Kemeny [1959] proposed a precise criterion for determining the "best" aggregate ranking. Given n candidates and k permutations of the candidates, $\pi_1, \ldots, \pi_k$, a *Kemeny-optimal* ranking of the candidates is a ranking π that minimizes the sum of *Kendall tau* distance between π and all π_i, $i = 1, \ldots, k$.[1] The specific formulation of rank aggregation where the goal is to find a Kemeny-optimal ranking has been often referred to as simply RANK-AGGREGATION in the literature.

RANK-AGGREGATION is an optimization problem where contradictory pieces of information are given as input and the goal is to find a globally consistent solution that minimizes the extent of disagreement with the respective inputs. As such, RANK-AGGREGATION is close in spirit to correlation clustering. A deeper connection between the two problems was firstly shown by Ailon et al. [2008a], who prove that (a slight variant of) the well-established Qwick-Cluster algorithm for correlation clustering (Algorithm 1.1, Section 1.5) can be used to solve RANK-AGGREGATION with similar quality guarantees.

Feedback arc set on tournaments.   A problem closely related to RANK-AGGREGATION—and, as such, to correlation clustering too—is the so-called *feedback arc set on tournaments* (FAS-TOURNAMENT, for short). A tournament is a directed graph where at least one arc exists between every pair of vertices. A minimum feedback arc set is the smallest set of arcs whose removal makes the graph become acyclic.

The relation between FAS-TOURNAMENT and RANK-AGGREGATION lies in the fact that the latter can be cast as a special case of a weighted variant of FAS-TOURNAMENT, where the

---

[1]The Kendall tau distance between two ranked lists π', π'' is defined as the number of pairs of candidates that are ranked in different orders by π' and π''.

objective is to minimize the total weight of backward edges in a linear order of the vertices. When the weight of arc (i, j) is the fraction of input rankings that order i before j, solving RANK-AGGREGATION is equivalent to solving this weighted FAS-TOURNAMENT instance [Ailon et al., 2008a].

1.9   APPLICATIONS

Correlation clustering has been employed in a large variety of application scenarios. In the following, we briefly overview some of the most relevant ones.

One of the earliest yet most natural applications of correlation clustering is in the context of *document clustering*, where the goal is to partition a given set of documents into topics [Bansal et al., 2004]. In this setting it is typically hard getting a precise idea of what a "topic" corresponds to. What is easier to obtain is instead a classifier—learned from historical data—that outputs whether or not it believes any two documents in the given input set are similar to each other. With such a classifier in place, one may apply it to every pair of input documents, and then find the clustering that agrees as much as possible with the classifier predictions.

Similar in spirit is the problem of *clustering entity names*, where items correspond to entries taken from multiple datasets (e.g., names or affiliations of researchers), and the goal is to collect together the entries that correspond to the same entity (person) [Cohen and Richman, 2002].

DasGupta et al. [2007] focus on the *decomposition of biological networks* into monotone subsystems. Specifically, given a network representing positive/negative interactions between proteins, RNA, DNA, metabolites, and other species, DasGupta et al. [2007] show how correlation clustering naturally applies to solve the fundamental problem of computing the smallest number of edges that have to be removed from the input network so that there remains a "sign-consistent" graph, i.e., a graph where all closed loops have an even number (possibly zero) of negative edges.

Correlation clustering has been also employed in other problems from the computational-biology domain, such as, e.g., *analysis of gene-expression data*, specifically in the detection of groups of genes that manifest similar expression pattern [Ben-Dor et al., 1999].

The databases community has applied correlation clustering to the problem of *duplicate detection*, also known as *entity resolution* or *record linkage*, which consists in identifying database records that potentially refer to the same real-world entity [Hassanzadeh et al., 2009].

Duplicate detection, but from a natural-language-processing (NLP) perspective, has been targeted by Finkel and Manning [2008]. Particularly, they devise a correlation-clustering-like formulation of the problem of *coreference resolution*, i.e., the task of deciding which noun phrases, or mentions, in a document refer to the same real-world entity.

Still in the NLP domain, Elsner and Charniak [2008] investigate the use of correlation clustering for *conversation disentanglement*, that is the problem of dividing a transcript into a set of distinct conversations.

In the context of *social* and *information networks*, Veldt et al. [2018] apply correlation clustering to the well-established problem of *community detection*, while Cesa-Bianchi et al. [2012] develop a theory of *link classification* in signed networks using a correlation-clustering-derived index as a measure of label regularity.

Correlation clustering has been largely employed in *computer vision*, specifically in the problem of *image segmentation*, where the goal is to decompose an image into a previously unknown number of segments that are somehow homogeneous but do not belong to a predefined set of categories [Björn et al., 2011, Kim et al., 2011, Yarkony et al., 2012]. In this context, Björn et al. [2011] devise an approach where an oversegmentation of the input image is first performed to represent the image as a graph with vertices corresponding to image regions and edges correspond to adjacent regions. Then, a binary random variable is associated with each edge, stating whether the corresponding curve should become active as part of a segment boundary, or remain dormant. Ultimately, a probabilistic graphical model is designed, which associates probabilities with the various realizations of random variables, and the solution to the image-segmentation problem is derived by properly partitioning the graph associated with the input image based on that probabilistic graphical model. This is shown to have connection with the multicut problem, and, as such, with correlation clustering too. Kim et al. [2011] present a formulation of correlation clustering that takes into account higher-order cluster relationships. This is motivated in the context of image segmentation as it improves clustering in the presence of local boundary ambiguities. Kim et al. [2011] first apply the pairwise correlation clustering to image segmentation over a pairwise superpixel graph and then develop higher-order correlation clustering over a hypergraph that considers higher-order relations among superpixels. Yarkony et al. [2012] define a optimization scheme for (weighted) correlation clustering that specifically exploits the planar structure of a graph derived from an image. Yarkony et al. [2012]'s method provides lower-bounds on the energy of the optimal correlation clustering that are typically fast to compute and tight in practice. Such a method is shown to outperform traditional global optimization techniques in minimizing the same objective.

Bonchi et al. [2019] study the problem of *discovering polarized communities in signed networks*, i.e., two communities (subsets of the network vertices) where within communities there are mostly positive edges while across communities there are mostly negative edges. They formulate their problem as a variant of correlation clustering where only two clusters are sought (see Section 2.1), while allowing some vertices not to be part of any cluster (see Section 3.3).

Mandaglio et al. [2020] show how correlation clustering can be exploited to solve clustering problems in which the pairwise interactions between entities are characterized by probability distributions and conditioned by external factors within the environment where the entities interact. In particular, Mandaglio et al. [2020] consider the case where the interaction conditioning factors can be modeled as cluster memberships of entities in a graph and the goal is to partition a set of entities such as to maximize the overall vertex interactions or, equivalently, minimize the loss of interactions in the graph.

Further applications of correlation clustering include [Pandove et al., 2018]: *aggregation tasks*, such as *rank aggregation* [Dwork et al., 2001] and *clustering aggregation* [Gionis et al., 2007]; *disambiguation in people search* [Kalashnikov et al., 2008]; and *web-search query clustering* [McCallum and Wellner, 2005], which in turn enables *automatic labeling of URL-query pairs* [Agrawal et al., 2009] and *segmentation of web pages* [Chakrabarti et al., 2008].

CHAPTER 2

Constraints

This chapter discusses alternative formulations of correlation clustering where *constraints* are added to the basic formulation. Although adding constraints results in a shrinkage of the feasible region, the corresponding constrained formulations can be interpreted as *generalizations* of the basic correlation-clustering formulation, as the latter can be obtained from a constrained formulation by setting the additional constraints to *ad hoc* values.

Specifically, this chapter focuses on formulations of correlation clustering where the number of output clusters is fixed (Section 2.1), where the size of the output clusters is bounded by some constant(s) (Section 2.2), and where the output clustering has a pre-specified error (Section 2.3).

2.1 CORRELATION CLUSTERING WITH A FIXED NUMBER OF CLUSTERS

As extensively remarked in Chapter 1, in correlation clustering the number of output clusters is automatically discovered and does not need to be specified as a user-defined parameter (as it is the case in most clustering methods). Although this feature is appealing in most contexts, fixing the number of clusters may still be required in some other contexts, e.g., whenever the application domain at hand enforces the fact that a good output partitioning should not be too large. For this reason, the variant of correlation clustering where the number of output clusters is constrained to be upper-bounded by some integer $k \geq 2$ has been well-studied in the literature [Bansal et al., 2004, Bonchi et al., 2013a, Coleman et al., 2008a, Giotis and Guruswami, 2006b, Il'Ev and Navrotskaya, 2016, Karpinski and Schudy, 2009].

We next present both the disagreement-minimization and agreement-maximization variants that have been considered in the literature. Here the input graph G is typically considered complete, i.e., with an edge between every (unordered) pair of vertices. We make this assumption throughout the rest of this section as well, apart from Section 2.1.4, where we discuss the case of general graphs.

Problem 2.1 (Min-Disagree[k]) Given an integer $k \geq 2$, and a signed graph $G = (V, E, \mathcal{L})$, where V is a set of n vertices, $E \subseteq \binom{V}{2}$ is a set of m edges, and $\mathcal{L} : E \rightarrow \{-1, +1\}$ is a function labeling an edge as either positive or negative, find a clustering $\ell : V \to \mathbb{N}$ so as to

$$\text{minimize} \sum_{\substack{(u,v)\in E, \\ \ell(u)=\ell(v)}} \frac{1 - \mathcal{L}(u,v)}{2} + \sum_{(u,v)\in E, \ell(u)\neq\ell(v)} \frac{1 + \mathcal{L}(u,v)}{2}$$

subject to $\quad |\{\ell(u) \mid u \in V\}| \leq k.$

Problem 2.2 (Max-Agree[k]) Given an integer $k \geq 2$, and a signed graph $G = (V, E, \mathcal{L})$, where V is a set of n vertices, $E \subseteq \binom{V}{2}$ is a set of m edges, and $\mathcal{L} : E \to \{-1, +1\}$ is a function labeling an edge as either positive or negative, find a clustering $\ell : V \to \mathbb{N}$ so as to

$$\text{maximize} \sum_{\substack{(u,v)\in E, \\ \ell(u)=\ell(v)}} \frac{1 + \mathcal{L}(u,v)}{2} + \sum_{\substack{(u,v)\in E, \\ cl(u)\neq\ell(v)}} \frac{1 - \mathcal{L}(u,v)}{2}$$

subject to $\quad |\{\ell(u) \mid u \in V\}| \leq k.$

The next section discusses hardness results, then the following sections present algorithmic results.

2.1.1 HARDNESS RESULTS

Despite **NP**-hardness has been shown for general (unbounded) correlation clustering, the involved reductions (e.g., the one of Bansal et al. [2004]) rely on the number of clusters growing with the input size, and thus they do not carry over to the case when the number of clusters is a fixed constant. For this reason, ad-hoc proofs are needed. In particular, as the two variants of the problem—MIN-DISAGREE[k] and MAX-AGREE[k]—have complementary objectives, it clearly suffices to prove **NP**-hardness of one of them to have **NP**-hardness of the other variant easily implied.

Shamir et al. [2004] originally provided a reduction for MAX-AGREE[2] from BALANCED-2-COLORING of a 3-uniform hypergraph, that is the problem of determining whether a 3-uniform hypergraph (i.e., a hypergraph whose hyperedges have all cardinality 3) admits a 2-coloring such that the number of vertices that are colored by each color is equal to each other [Lovasz, 1973]. Then, they show that such a result can be generalized to MAX-AGREE[k], for any $k > 2$, via a reduction from MAX-AGREE[2], where a MAX-AGREE[k]'s problem instance is constructed by adding $k - 2$ sets of vertices of size n^2 each to original MAX-AGREE[2]'s instance, and drawing a positive edge between vertices u and v if and only if they belong to the same of such added sets.

An alternative yet simpler proof is devised by Giotis and Guruswami [2006b]. Specifically, they show a reduction for MIN-DISAGREE[2] from GRAPH-MIN-BISECTION, that is the problem of partitioning the vertex set of a graph into two equal halves so that the number of edges connecting vertices in different halves is minimized, whose **NP**-hardness is known since the 1970s (see, e.g., Theorem 1.3 in Garey et al. [1976]). Then, they generalize this result to the case $k > 2$ in a way analogous to Shamir et al. [2004]. The details of Giotis and Guruswami [2006b]'s **NP**-hardness proof are as follows.

Theorem 2.3 [**Giotis and Guruswami, 2006b**].   MIN-DISAGREE[2] *is* **NP**-*hard.*

Proof. Given an instance (i.e., a simple, undirected graph) $G' = (V', E')$ of GRAPH-MIN-BISECTION (with an even number of vertices), an instance $G = (V, E, \mathcal{L})$ of MIN-DISAGREE[2] is constructed as follows. First, let $V = V' \cup \bigcup_{x \in V'} V_x$, where, for every vertex $u \in V'$, V_x is a set of additional vertices of size $|V_u| = |V'|$. Call any $\{u\} \cup V_u$ "group." As far as the labeling of the edges, set all edges in E' to be positive edges, and all edges not in E' to be negative. Moreover, label all edges between vertices in the same group as positive, and all other edges (i.e., edges between vertices within different groups, or between a vertex in V and a vertex in some group) as negative. Such a construction clearly takes polynomial time in the size of G'.

The proof is based on showing that any 2-clustering of G with the minimum number of disagreements has two clusters of equal size with all vertices of any group in the same cluster. Consider some optimal 2-clustering W with two clusters W_1 and W_2 such that $|W_1| \neq |W_2|$ or not all vertices of some group are in the same cluster. Pick some group V_u such that not all its vertices are put in the same cluster. Place all the vertices of the group in the same cluster, obtaining W' such that the size difference of the two clusters is minimized. If such a group could not be found, pick a group V_u from the larger cluster and place it in the other cluster. Since all groups contain the same number of vertices, it must be the case that the size difference between the two clusters is reduced.

Assume that V_u^1 vertices of group V_u were in W_1 and V_u^2 in W_2. Without loss of generality, assume also that the clustering $W' = (W_1', W_2')$ above is obtained by moving the V_u^1 group vertices into cluster W_2, i.e., $W_1' = W_1 \setminus V_u^1$, $W_2' = W_2 \cup V_u^1$. The following facts about the difference in the number of disagreements between W' and W can be observed:

- Clearly, the number of disagreements between vertices not in V_u and between one vertex in V_u^2 with one in W_1' remains the same.

- The number of disagreements is decreased by $|V_u^1| \times |V_u^2|$, based on the fact that all those edges are positive.

- It is also decreased by at least $|V_u^1| \times |W_1'| - (|V'| - 1)$, based on the fact that all but at most $|V'| - 1$ edges between a vertex in V_u and any other vertex not in V_u are negative.

- The number of disagreements increases at most $|V_u^1| \times |W_2 \setminus V_u^2|$, because (possibly) all vertices in V_u^1 share a negative edge with vertices in W_2 outside their group.

Overall, the difference in the number of disagreements is at most:

$$|V_u^1| \times |W_2 \setminus V_u^2| - |V_u^1| \times |V_u^2| - |V_u^1| \times |W_1'| + (|V'| - 1).$$

Since $||W_1'| - |W_2'||$ was minimized, it must be the case that $|W_1'| \geq |W_2 \setminus V_u^2|$. Moreover, since a group has an odd number of vertices and the total number of vertices of V is even, it follows that $|W_1'| \neq |W_2 \setminus V_u^2|$, and thus $|W_1'| - |W_2 \setminus V_u^2| \geq 1$. Therefore, the total number of disagreements increases at most $(|V'| - 1) - |V_u^1| \times (|V_u^2| + 1)$. Since $|V_u^1| + |V_u^2| = |V'| + 1$ and V_u^1 cannot be empty, it follows that $|V_u^1| \times (|V_u^2| + 1) \geq |V'|$ and the number of disagreements strictly decreases, contradicting the optimality of W. This implies that the optimal solution to the MIN-DISAGREE[2] instance has two clusters of equal size and all vertices of any group are contained in a single cluster. It is now trivial to see that an optimal solution to the GRAPH-MIN-BISECTION problem can be easily derived from the MIN-DISAGREE[2] solution by simply discarding the vertices not in V'. This completes the reduction.   □

Theorem 2.4   [Giotis and Guruswami, 2006b].   MIN-DISAGREE[k] *is* **NP**-*hard, for every* $k \geq$ 2.

Proof. Given an instance $G = (V, E, \mathcal{L})$ of MIN-DISAGREE[2], create an augmented graph $G' = (V', E')$, where V' is computed by adding to V a number of $k - 2$ groups of $n + 1$ vertices each, and adding a positive edge between vertices within each of such added groups and a negative edge between vertices in different groups. Consider now a k-clustering of G' such that the number of disagreements is minimized. The theorem follows from the fact that such a k-clustering contains two clusters that are composed of all and only vertices in V, and these two clusters are an optimal solution for original MIN-DISAGREE[2]'s problem instance. To this end, observe that all the vertices of a group must make up one cluster. Also, any of the original vertices cannot end up in one group's cluster since that would induce $n + 1$ disagreements, strictly more than it could possibly induce in any of the two remaining clusters. Therefore, the two non-group clusters are an optimal 2-clustering for MIN-DISAGREE[2] on G.   □

2.1.2   POLYNOMIAL-TIME APPROXIMATION SCHEMES

The most relevant yet general approximation results for correlation clustering with a fixed number of clusters are due to Giotis and Guruswami [2006b] and Karpinski and Schudy [2009], who show that both MIN-DISAGREE[k] and MAX-AGREE[k] admit a PTAS, for every $k \geq 2$. All these PTASs are randomized: this means that they deliver a solution with the claimed approximation guarantee with high probability. Next, we discuss such algorithms in more detail.

A PTAS for Max-Agree[k]

Giotis and Guruswami [2006b] devise a PTAS for MAX-AGREE[k] which uses random sampling and follows closely the property tester for the MAX-k-CUT problem devised by Goldreich et al. [1998].[1] The basic observation of Giotis and Guruswami [2006b] is that for every $k \geq 2$, and every instance of MAX-AGREE[k], the optimal number OPT of agreements is at least $n^2/16$.

Lemma 2.5 **[Giotis and Guruswami, 2006b]**   *For every instance $\langle G, k \rangle$ of* MAX-AGREE[k], *the optimal objective-function value OPT is $\geq n^2/16$.*

Proof. Let $n_+ = |\{(u, v) \in E \mid \mathcal{L}(u, v) = +1\}|$ and $n_- = \binom{n}{2} - n_+$. By placing all vertices in a single cluster, we get n_+ agreements. By placing objects randomly in one of k clusters, we get an expected $(1 - 1/k)n_-$ agreements just on vertex pairs connected by a negative edge. Therefore:

$$\mathsf{OPT} \; \geq \; \max\{n_+, (1 - 1/k)n_-\} \; \geq \; \frac{(1 - 1/k)}{2}\binom{n}{2} \; \geq \; \frac{n^2}{16}.$$

$\square$

With Lemma 2.5 in place, to ultimately have a PTAS, it suffices to devise an algorithm that guarantees a solution within additive $\mathcal{O}(\varepsilon n^2)$ of OPT for arbitrary ε. Giotis and Guruswami [2006b] devise such an algorithm and term it MaxAg(k, ϵ). The algorithm is outlined in Algorithm 2.3, and described next.

MaxAg(k, ϵ) works in $s = \mathcal{O}(1/\varepsilon)$ steps. The input vertices are partitioned into s almost equal-size sets $V^1, \ldots, V^s$ (each with $\Theta(\varepsilon n)$ vertices) in an arbitrary way. In the i-th step, vertices within V^i are clustered, with the aid of a sufficiently large, but constant-sized, random sample S^i drawn from vertices outside V^i, and with the random samples for different steps chosen independently. The algorithm then tries all possible ways in which all the samples can be clustered, and for each possibility, it clusters V^i by placing every one of its vertices in the cluster that maximizes the agreement with respect to the clustering of the S^i sample. The approximation guarantee of MaxAg(k, ϵ) is as follows.

Theorem 2.6 **[Giotis and Guruswami, 2006b].**   *Given an instance $\langle G, k \rangle$ of* MAX-AGREE[k], *and real numbers $\varepsilon > 0$, $\delta \in (0, 1)$, with probability at least $1 - \delta$, the MaxAg(k, ϵ) algorithm outputs a k-clustering whose objective-function value is at least $\mathsf{OPT} - \varepsilon n^2/2$, where OPT is the optimal objective-function value of* MAX-AGREE[k] *on the instance $\langle G, k \rangle$. The running time of the algorithm is $nk^{\mathcal{O}(\varepsilon^{-3} \log(k/(\varepsilon\delta)))}$.*

Proof. (sketch) The analysis is based on the following rationale. Denote an optimal clustering by $\ell(\mathsf{OPT})$. Assume that the considered clustering of S^i matches the clustering of $V^1 \cup \cdots \cup V^{i-1}$

---

[1]MAX-k-CUT is the problem of partitioning the vertex set of a graph into k sets so as to maximize the sum of the number of edges between vertices in different sets.

Algorithm 2.3 MaxAg(k, ϵ) [Giotis and Guruswami, 2006b]

Input: An instance $\langle G, k \rangle$ of MAX-AGREE[k], $\varepsilon > 0$, $\delta \in (0, 1)$
Output: A k-clustering $\ell : V \to \mathbb{N}$
  1: Construct an arbitrary partition of V into roughly equal parts, $(V^1, V^2, \ldots, V^s)$, $s = \lceil \frac{4}{\varepsilon} \rceil$
  2: **for** $i = 1, \ldots s$ **do**
  3:     Choose uniformly at random with replacement from $V \setminus V^i$ a subset S^i of size $r = \frac{32^2}{2\varepsilon^2} \log \frac{64sk}{\varepsilon\delta}$
  4: **end for**
  5: Set ℓ to be an arbitrary (or random) clustering
  6: **for all** clusterings of each of the sets S^i into $(S_1^i, \ldots, S_k^i)$ **do**
  7:     **for** $i = 1, \ldots, s$ and every vertex $u \in V^i$ **do**
  8:         $\forall j = 1, \ldots, k$, let $\beta_j(u) = |N^+(u) \cap S_j^i| + \sum_{l \neq j} |N^-(u) \cap S_l^i|$
  9:         Place u in cluster $\text{argmax}_j \beta_j(u)$
  10:     **end for**
  11:     If the clustering formed by Steps 7–10 has more agreements than ℓ, set ℓ to be that clustering
  12: **end for**

---

the algorithm has computed so far and the optimal clustering on $V^{i+1} \cup \cdots \cup V^s$ (since the algorithm tries all possible clusterings of S^i, this particular clustering will be tried as well). In this case, using standard random-sampling bounds, one can show that, with high probability over the choice of S^i, the clustering of V^i chosen by the algorithm is quite close, in terms of number of agreements, to the clustering of V^i by the optimal clustering $\ell(\text{OPT})$. This implies that with constant probability the choice of the algorithm on S^i allows us to place vertices in such a way that the decrease in the number of agreements with respect to an optimal clustering is $\mathcal{O}(\varepsilon^2 n^2)$. Thus, the algorithm is guaranteed to output a solution that is at most $\mathcal{O}(\varepsilon n^2)$ fewer agreements compared to the optimal solution.   □

It is interesting to note that the MaxAg(k, ϵ) algorithm can also be used as a PTAS for the unbounded-clusters case, by setting $k = \Omega(1/\varepsilon)$. It is then not hard to see by a cluster-merging procedure that the solution obtained by MaxAg(k, ϵ) is still within an ε-factor of the optimal solution. Furthermore, since MaxAg(k, ϵ) runs in linear time in the number of input vertices, it can be a more efficient alternative to the PTAS presented in Bansal et al. [2004], whose running time is $\mathcal{O}(n^2 e^{\mathcal{O}(\frac{1}{\varepsilon^{10}} \log(\frac{1}{\varepsilon}))})$.

PTASs for Min-Disagree[k]

The most relevant result for MIN-DISAGREE[k] is the existence of a PTAS [Giotis and Guruswami, 2006b, Karpinski and Schudy, 2009]. This is a quite surprising finding in light of the APX-hardness of the corresponding version of the problem when the number k of clusters is not specified to be a constant (recall that the maximization version does instead admit a PTAS even when k is not specified). However, as remarked by Giotis and Guruswami [2006b], it is not uncommon that minimization versions of problems are harder compared to their complementary maximization versions. The APX-hardness of MIN-DISAGREE despite the existence of a PTAS for MAX-AGREE is a notable example. A major difficulty in those cases is to handle the instances where the optimal value of the minimization version is very small. For those instances, in fact, even a PTAS for the complementary maximization problem may easily fail in providing a good approximation for the minimization problem.

The first PTAS for MIN-DISAGREE[k] is proposed by Giotis and Guruswami [2006b]. It is termed MinDisag(k, ϵ) and uses MaxAg(k, ϵ) (Algorithm 2.3) as a subroutine. The MinDisag(k, ϵ) PTAS is more elaborated and comes with a much more complicated analysis than the PTAS for the maximization counterpart. For this reason, we defer to Giotis and Guruswami [2006b] for all technical details. Here we just report the most significant high-level remarks. In this respect, Giotis and Guruswami [2006b] point out that the difficulty in obtaining a PTAS for the minimization version of the problem is similar to that faced in the problem of MIN-k-SUM-CLUSTERING, which has the complementary objective function to METRIC-MAX-k-CUT.[2] In fact, although the case of MIN-2-SUM-CLUSTERING was solved in Indyk [1999] soon after the algorithm for METRIC-MAX-k-CUT by de la Vega and Kenyon [1998], the case $k > 2$ appeared harder. Similarly to this, for MIN-DISAGREE[k], it is quite easy to give a PTAS for the 2-clustering version using the MaxAg(k, ϵ) algorithm for MAX-AGREE[k], but much more effort is needed for the case of $k > 2$ clusters. Another relevant observation is that the PTAS for MAX-AGREE[k] would in principle provide a very good approximation for general (i.e., $k \geq 2$) MIN-DISAGREE[k] too, but only when the optimal number of disagreements is not too small. Thus, the hardest work in the design and analysis of the MinDisag(k, ϵ) PTAS is for the case when the optimal clustering has very few disagreements.

Ultimately, the result provided in Giotis and Guruswami [2006b] is as follows.

Theorem 2.7 [Giotis and Guruswami, 2006b].   *For every $k \geq 2$, there is a (randomized) PTAS for MIN-DISAGREE[k], with running time $n^{\mathcal{O}(9^k/\varepsilon^2)} \log n$.*

Karpinski and Schudy [2009] notice that MIN-DISAGREE[k] can be modeled as a *dense constraint satisfaction problem* (CSP), and, based on this, devise an alternative PTAS with running time better than Giotis and Guruswami [2006b]'s one.

A CSP consists of N variables, each of which taking values from a domain of size D, and a collection of arity-K constraints (both D and K constant). Every constraint is a pair consisting of a subset of variables and a relation on the values assigned to those variables. A CSP may be modeled as either a minimization or a maximization problem: the objective of MIN-K-CSP

---

[2]MIN-k-SUM-CLUSTERING is the problem of partitioning a set of points endowed with a metric into k clusters so as to minimize the sum of the distances between points in the same cluster. METRIC-MAX-k-CUT is the problem of dividing a set of points in metric space into two parts so as to maximize the sum of the distances between points belonging to distinct parts.

(resp. Max-K-CSP) is to assign a value—from the given domain—to each of the N variables so as to minimize the number of unsatisfied (resp. maximize the number of satisfied) constraints. An (everywhere) *dense* instance of a CSP is one where every variable is involved in at least a constant times the maximum possible number of constraints, i.e., $\Omega(n^{K-1})$. A constraint is *fragile* if modifying any variable in a satisfied constraint makes the constraint unsatisfied. A CSP is *fragile* if all of its constraints are.

Min-Disagree[k] can be modeled as a dense Min-2-CSP, where every vertex in the input graph is assigned a variable (thus, $N = n$), the k cluster labels constitute the domain (thus, $D = k$), and a constraint for every edge is defined, requiring that the values (cluster labels) assigned to the variables of the vertices of a positive (resp. negative) edge must be equal (resp. different). Min-Disagree[2] is also fragile, and the following result, holding in general for dense fragile Min-2-CSP, applies to it.

Theorem 2.8   [Karpinski and Schudy, 2009].   *There is a (randomized) PTAS for* Min-Disagree[2] *with running time* $\mathcal{O}(n^2) + 2^{\mathcal{O}(1/\varepsilon^2)}$.

For $k > 2$, Min-Disagree[k] is not fragile, but still has properties allowing for a PTAS anyway. Specifically, the following holds.

Theorem 2.9   [Karpinski and Schudy, 2009].   *For every $k > 2$, there is a (randomized) PTAS for* Min-Disagree[k] *with running time* $n^2 2^{\mathcal{O}(k^6/\varepsilon^2)}$.

Such a result improves on the running time of Giotis and Guruswami [2006b]'s PTAS in two ways: first, the exponent of the input-dependent term is constant rather than dependent on ε, and, second, the ε-dependent term is polynomial in k rather than exponential.

2.1.3   OTHER APPROXIMATION RESULTS AND ALGORITHMS

The very first approximation result for correlation clustering with a fixed number of clusters is about the special case $k = 2$ [Bansal et al., 2004]. Specifically, Bansal et al. [2004] devise the following simple 3-approximation algorithm for Min-Disagree[2], which, borrowing the terminology in Coleman et al. [2008a], we call Pick-a-Vertex. For every input vertex $u \in V$, let

$$N^+(u) = \{v \in V \mid \mathcal{L}(u,v) = +1\},$$

$$N^-(u) = \{v \in V \mid \mathcal{L}(u,v) = -1\}.$$

Bansal et al. [2004]'s Pick-a-Vertex algorithm simply considers all 2-clusterings $\ell_u = \{\{u\} \cup N^+(u), N^-(u)\}$, for all $u \in V$, and, among all those clusterings ℓ_u, outputs the one that minimizes the number of disagreements. The following holds.

Theorem 2.10   [Bansal et al., 2004].   *Pick-a-Vertex is a 3-approximation algorithm for* Min-Disagree[2].

Proof. Denote by $\ell^* = \{C_1^*, C_2^*\}$ the optimal solution of Min-Disagree[2] on the given input instance, and by $dis(\ell^*)$ the number of disagreements of ℓ^*.

Call an edge "bad" if ℓ^* disagrees with it, and define the "bad degree" of a vertex to be the number of bad edges incident to it. Clearly, if there is a vertex that has no bad edges incident to it, the clustering produced by that vertex would be the same as $\{C_1^*, C_2^*\}$, and it would have as many disagreements as $dis(\ell^*)$.

Otherwise, let u be a vertex with minimum bad degree d, and, without loss of generality, let $u \in C_1^*$. Consider the clustering $\ell_u = \{\{u\} \cup N^+(u), N^-(u)\}$. Let X be the set of bad neighbors of u, i.e., the d vertices that are in wrong ℓ_u's cluster with respect to $\{C_1^*, C_2^*\}$. The total number of extra disagreements due to this set X (other than the disagreements already made by ℓ^*) is at most nd. However, since all vertices have bad degree at least d, then it holds that $dis(\ell^*) \geq nd/2$. Hence, the number of extra disagreements made by clustering ℓ_u is at most $2dis(\ell^*)$. $\qquad\square$

Coleman et al. [2008a] remark that the 3-approximation factor of the Pick-a-Vertex algorithm is tight. To this purpose, consider a complete graph consisting of solely positive edges, apart from a Hamiltonian cycle of negative edges. In that instance, the optimal solution, i.e., placing all vertices in the same cluster, has cost n, while the output of Pick-a-Vertex has cost $3n - 10$.

Coleman et al. [2008a] also notice that Pick-a-Vertex yields solutions that are not locally optimal. Based on this observation, they devise a local-search algorithm for Min-Disagree[2] that achieves a factor-2 approximation guarantees. The algorithm is termed PASTA-toss (where "PASTA" comes from "Pick-a-Spanning-Tree algorithm"), and follows exactly the Pick-a-Vertex algorithm, apart from the fact that a local-search procedure is run on every clustering $\ell_u = \{\{u\} \cup N^+(u), N^-(u)\}$ to yield the ultimate candidate clusterings. Such a local search is a standard one: the best (in terms of disagreement minimization) swap of a vertex from one of the two current clusters to the other one is iteratively performed, until there exist no swaps leading to a decrease in the number of disagreements of the current clustering. Call ℓ_u' the clustering obtained by running local search on ℓ_u. All the ℓ_u' are by-design guaranteed to be local optima. Among all such ℓ_u', the PASTA-toss algorithm ultimately outputs the one achieving the smallest number of disagreements. The following result is provided in Coleman et al. [2008a]. The proof is rather articulated and is thus omitted.

Theorem 2.11   [Coleman et al., 2008a].   *PASTA-toss is a 2–approximation algorithm for* Min-Disagree[2].

Il'Ev and Navrotskaya [2016] extend Coleman et al. [2008a]'s PASTA-toss algorithm to devise a further local-search algorithm that is shown to provide a k-factor approximation guarantee for Min-Disagree[k], for any $k \geq 2$.

Bonchi et al. [2019] study a relaxation of MIN-DISAGREE[2] where not necessarily all input vertices should be part of a clustering, but some of them may be identified as outliers and left apart.

2.1.4   GENERAL GRAPHS

All the discussions reported so far refer to the case where the input graph is complete, i.e., there is an edge between every (unordered) pair of vertices. In the following, we complement the discussion by focusing on the general-graph case. Specifically, we first report some results due to Giotis and Guruswami [2006b] on the complexity of MIN-DISAGREE[k] and MAX-AGREE[k] when the input graph may contain an arbitrary number of edges. It will become clear from the reported results—which are consequences of connections to problems like MAX-CUT and graph colorability—that the general-graph case is much harder than the complete-graph one. Then, we also discuss heuristic and exact algorithms for MIN-DISAGREE[2] devised by Coleman et al. [2008a] and Hüffner et al. [2007], respectively.

Approximation Results for Min-Disagree[k] and Max-Agree[k]

Theorem 2.12   [Giotis and Guruswami, 2006b]. *There is a polynomial-time 0.878-approximation algorithm for* MAX-AGREE[2] *on general graphs. For every k > 2, there is a polynomial-time 0.7666-approximation algorithm for* MAX-AGREE[k] *on general graphs.*

Proof. The bound for the 2-clustering case follows from a connection with the well-established MAX-CUT problem, where an easy trick to account for positive edges is employed. The classic SDP-based 0.878-approximation algorithm for MAX-CUT by Goemans and Williamson [1995] may be therefore adapted to achieve the same approximation guarantee on MAX-AGREE[2] too. Specifically, the idea is to use a semidefinite program relaxation where every vertex is assigned a unit vector, and the objective function, which now includes terms for the positive edges, equals

$$\sum_{\substack{(u,v)\in E, \\ \mathcal{L}(u,v)=-1}} \frac{1 - \langle \vec{v}_u, \vec{v}_v \rangle}{2} + \sum_{\substack{(u,v)\in E, \\ \mathcal{L}(u,v)=+1}} \frac{1 + \langle \vec{v}_u, \vec{v}_v \rangle}{2}.$$

The rounding is identical to Goemans and Williamson [1995]'s random-hyperplane one. By the analysis in Goemans and Williamson [1995], it holds that the probability that $\vec{v}_u$ and $\vec{v}_v$ are separated by a random hyperplane is at least 0.878 times $(1/2)(1 - \langle \vec{v}_u, \vec{v}_v \rangle)$. By a similar calculation, it can be shown that the probability that $\vec{v}_u$ and $\vec{v}_v$ are not separated by a random hyperplane is at least 0.878 times $(1/2)(1 + \langle \vec{v}_u, \vec{v}_v \rangle)$. These facts imply that the expected agreement of the clustering produced by random-hyperplane rounding is at least 0.878 times the optimal value of the above semidefinite program, which in turn is at least as large as the maximum agreement with two clusters.

The bound for $k > 2$ is obtained by Swamy [2004], who also notes that slightly better bounds are possible for $2 < k \leq 5$. $\qquad\square$

Giotis and Guruswami [2006b] also remark that, due to the connection with MAX-CUT, and the inapproximability result for MAX-CUT by Khot et al. [2007], the above approximation guarantee for MAX-AGREE[2] on general graphs is the best possible, unless the *unique games conjecture* [Khot, 2002] is false.

Theorem 2.13   **[Giotis and Guruswami, 2006b].**   *There is a polynomial-time $\mathcal{O}(\sqrt{\log n})$-approximation algorithm for* MIN-DISAGREE[2] *on general graphs. For $k > 2$,* MIN-DISAGREE[k] *on general graphs cannot be approximated within any finite factor.*

Proof. The bound for 2-clustering follows by the simple observation that MIN-DISAGREE[2] on general graphs reduces to MIN-2CNF-DELETION, i.e., given an instance of 2SAT, determining the minimum number of clauses that have to be deleted to make it satisfiable. The latter problem admits an $\mathcal{O}(\sqrt{\log n})$-approximation algorithm [Agarwal et al., 2005].

The result on MIN-DISAGREE[k] for $k > 2$ follows by a reduction from k-coloring. When $k > 2$, it is **NP**-hard to tell if a graph is k-colorable, and thus even given an instance of MIN-DISAGREE[k] with only negative edges, it is **NP**-hard to determine if the optimum number of disagreements is zero or positive. $\qquad\square$

Heuristics for Min-Disagree[2]

A number of heuristics have been also proposed for MIN-DISAGREE[2] on general graphs [Coleman et al., 2008a,b]. Two of such heuristics are variants of the Pick-a-Vertex and PASTA-toss algorithms introduced above. Recall that the first step of both Pick-a-Vertex and PASTA-toss consists in yielding a set of n candidate 2-clusterings $\{\ell_u = \{u \cup N^+(u), N^-(u)\}\}_{u \in V}$. In complete graphs ℓ_u is clearly guaranteed to correspond to a partition of the input vertex set, while in general graphs this is not necessarily the case, as some vertices may be left apart. For this reason, the idea behind the heuristics devised in Coleman et al. [2008a] is simply to define every ℓ_u as the 2-clustering induced by the (rooted) spanning tree T_u produced by a breadth-first search originated in u. More specifically, the 2-clustering induced by some T_u is defined by putting the root vertex u in one of the two clusters arbitrarily (without loss of generality), visiting the rest of T_u in a breadth-first-search fashion, and assign every encountered vertex v to the cluster of v's father f_v if edge (v, f_v) is positive, and to the other cluster, otherwise.

Coleman et al. [2008a] also devise a further variant of PASTA-toss, termed, PASTA-flip, where a more elaborated local search is employed. Rather than simply tossing vertices between clusters, PASTA-flip flips edges (from positive to negative, or vice versa) that are involved in many "bad" cycles, i.e., cycles having an odd number of negative edges. The rationale is that bad cycles necessarily induce some disagreements in every output clustering, including the optimal one. Thus, flipping edges involved in many bad cycles would make those many bad cycles become good, and a clustering ultimately computed out of this flipping procedure would intuitively have a limited number of disagreements. More in detail, given a spanning tree T_u (resulting from a breadth-first search originated in vertex u), PASTA-flip does the following. While there is an edge $e \in T_u$ that is involved in more bad cycles than good ones, flip e. When there are no more such edges, flip edges that are not part of T_u. When there are no more edges to be flipped, recompute T_u on the graph resulting from the flipping process, and use the 2-clustering induced by such a recomputed T_u as a candidate clustering. Ultimately, among all candidate clusterings, output the one minimizing the number of disagreements.

Coleman et al. [2008b] show a connection with NORMALIZED-CUT, and exploit it to formulate MIN-DISAGREE[2] as an eigendecomposition problem, similar to spectral clustering. Based on that, they introduce algorithms employing relaxation techniques (either SDP or spectral) similar to the ones traditionally used for NORMALIZED-CUT.

An Exact Algorithm

Hüffner et al. [2007] devise an exact algorithm to solve the decision version of MIN-DISAGREE[2][3]: given a signed graph $G = (V, E, \mathcal{L})$ and an integer $\omega \in [0, m]$, is there a 2-clustering of the vertices of G whose number of disagreements is no more than ω? The idea behind the design of such an exact algorithm is to reduce MIN-DISAGREE[2] to EDGE-BIPARTIZATION, that is the problem of deciding whether a graph can be made bipartite by deleting a number of edges no more than a given integer ω. The reduction is rather simple: just replace every positive edge (u, v) of original MIN-DISAGREE[2]'s instance with two edges, (u, z_{uv}) and (z_{uv}, v), where z_{uv} is a newly added vertex.

Hüffner et al. [2007]'s algorithm combines the traditional, fixed-parameter, $\mathcal{O}(2^\omega m^2)$-time algorithm for EDGE-BIPARTIZATION by Guo et al. [2006] with some sophisticated data-reduction techniques.[4] This way, Hüffner et al. [2007] are able to obtain an algorithm that, while still being exponential-time, runs in practice much faster than a naïve brute-force one.

We conclude this section by pointing out that there exist additional results on correlation clustering with a fixed number of clusters, concerning bipartite graphs and the local setting. Such results are discussed elsewhere in the book, specifically in the chapters dedicated to those contexts (i.e., Sections 4.1 and 5.1, respectively).

2.2   CORRELATION CLUSTERING WITH CONSTRAINED CLUSTER SIZES

Puleo and Milenkovic [2015] introduce a generalization of correlation clustering where bounds on the size of the clusters are allowed. A constraint on the maximum size of a cluster may indeed

---

[3]Being MIN-DISAGREE[k] and MAX-AGREE[k] equivalent in terms of optimum, Hüffner et al. [2007]'s algorithm can optimally solve MAX-AGREE[2] as well.

[4]Guo et al. [2006]'s algorithm was very recently improved by Pilipczuk et al. [2019], who devise an algorithm running in $\mathcal{O}(1.077^\omega nm)$ time.

be needed in several applications, such as, e.g., community detection, where it is not uncommon to have a-priori information about the largest possible size of a community [Newman, 2006].

The problem studied by Puleo and Milenkovic [2015] focuses on the MIN-DISAGREE *variant* of correlation clustering, and takes *complete, edge-weighted graphs* as input. In particular, as a further noteworthy contribution, Puleo and Milenkovic [2015] handle more general weighted graphs than previously done in the literature: they resort to the general formulation of correlation clustering originally introduced by Ailon et al. [2008a] (Problem 1.6, Section 1.4), and consider more general constraints within that formulation.

In more detail, Ailon et al. [2008a]'s formulation of correlation clustering assumes that each edge e is assigned two nonnegative weights, w_e^+ and w_e^-, and a clustering incurs cost w_e^+ if e is placed between clusters, and incurs cost w_e^- if e is placed within a cluster. If no restrictions are imposed on the weights w_e^+ and w_e^-, then it may happen that $w_e^+ = w_e^- = 0$. This is clearly equivalent to model the case that edge e is absent, meaning that the input graph is a general one.

Due to the hardness of approximating correlation clustering on general graphs,[5] several constraints on the weights assigned to the various edges have been considered, in order to make the problem more tractable from an approximation point of view. These include, e.g., the probability constraint (i.e., $w_e^+ + w_e^- = 1$, for every edge e) [Ailon et al., 2008a, Bansal et al., 2004, Charikar et al., 2005], or the probability constraint together with the triangle-inequality restriction (i.e., $w_{uv}^- \leq w_{uz}^- + w_{zv}^-$, for all triples u, v, z of vertices) [Ailon et al., 2008a, Gionis et al., 2007]. The main contribution by Puleo and Milenkovic [2015] in this regard is to go beyond all those constraints. Specifically, Puleo and Milenkovic [2015] deal with the following more general constraints:

$$w_e^+ \leq 1, \text{ for every edge } e, \text{ and}$$
$$w_e^- \leq \tau, \text{ for every edge } e, \text{ for some } \tau \in [1, \infty), \text{ and} \qquad (2.1)$$
$$w_e^+ + w_e^- \geq 1, \text{ for every edge } e.$$

Such constraints include, among others, the probability constraint (obtainable setting $\tau = 1$).

As far as the bounds on the cluster sizes, Puleo and Milenkovic [2015] accommodate both a *hard constraint* on the size of every cluster, as well as a *soft constraint* that imposes a penalty on the objective for oversized clusters, without outright forbidding them. Specifically, to model the soft constraint, every cluster is allowed to contain at most $K + 1$ vertices, where K is a user-specified nonnegative integer. Moreover, each vertex u is assigned a "penalty" parameter φ_u (which is, again, provided as input by the user). If u is placed in a cluster C with more than $K + 1$

---

[5]MIN-DISAGREE on general graphs is known to be APX-hard, and the best known approximation ratio is $\mathcal{O}(\log n)$, due to Charikar et al. [2005] and Demaine et al. [2006] (see Section 1.5.3). Given the equivalence between the MIN-MULTICUT problem and MIN-DISAGREE on general graphs [Charikar et al., 2005, Demaine et al., 2006], any improvement on the approximation of the latter would also result in an improvement on the approximation of MIN-MULTICUT. This is unlikely since, assuming the Unique Games Conjecture [Khot, 2002], no constant-factor approximation algorithm may exist for MIN-MULTICUT [Chawla et al., 2006].

vertices, then the cost function is charged an additional penalty of $\varphi_u(|C| - (K + 1))$. Since this penalty is assessed separately for each vertex in an oversized cluster, the total penalty cost for a cluster C with $|C| > K + 1$ is $(|C| - (K + 1)) \sum_{u \in C} \varphi_u$. If instead a hard constraint is desired, it suffices to set $\varphi_u = 1$, for every vertex u. This way, in fact, any clusters in the resulting solution which are too large can then be split arbitrarily into clusters of size $K + 1$ and a "remainder cluster," yielding no net increase in cost, since it is assumed that $w_e^+ \leq 1$, for every edge e. Similarly, the case where no size constraint at all is desired can be handled by simply setting $\varphi_u = 0$, for every vertex u. Formally, the problem studied by Puleo and Milenkovic [2015] is as follows.

Problem 2.14 (Cluster-Size-Bounded-Correlation-Clustering)   Given a weighted graph $G = (V, E, w^+, w^-)$, where V is a set of n vertices, $E \subseteq \binom{V}{2}$ is a set of m edges, and $w^+ : E \to \mathbb{R}_0^+$, $w^- : E \to \mathbb{R}_0^+$ are two functions assigning nonnegative weights to each edge and satisfying Equation (2.1), a function $\varphi : V \to \mathbb{R}_0^+$ assigning a nonnegative penalty to each vertex in G, and a nonnegative integer K, find a clustering $\ell : V \to \mathbb{N}$ so as to minimize

$$\sum_{\substack{(u,v) \in E, \\ \ell(u) = \ell(v)}} w_{uv}^- + \sum_{\substack{(u,v) \in E, \\ cl(u) \neq \ell(v)}} w_{uv}^+ + \sum_{\substack{u \in V, \\ |\ell(u)| > K+1}} \varphi_u(|\ell(u)| - (K + 1)), \qquad (2.2)$$

where $|\ell(u)|$ denotes the size of the cluster of u, and w_{uv}^+, w_{uv}^-, φ_u are short forms of $w^+(u, v)$, $w^-(u, v)$, $\varphi(u)$, respectively.

Puleo and Milenkovic [2015] devise two approximation algorithms for Problem 2.14: an algorithm based on the region-growing technique, and a (randomized) pivoting algorithm. The details of these two algorithms are reported next.

2.2.1   A REGION-GROWING ALGORITHM

A linear-programming formulation.   The first algorithm devised by Puleo and Milenkovic [2015] for Problem 2.14 is a region-growing-based one that is inspired by the 4-approximation algorithm presented in Charikar et al. [2005] for MIN-DISAGREE. In this regard, Puleo and Milenkovic [2015] preliminarily observe that Charikar et al. [2005]'s algorithm, although being originally conceived for unweighted graphs, works without modifications and achieves the same factor-4 approximation for weighted graphs with weights obeying the probability constraint.

In order to handle the aforementioned more general constraints on the edge weights (Equation (2.1)), instead, a number of changes are required, both in the linear-programming formulation at the basis of the algorithm, and the region-growing procedure itself. Specifically, the linear-programming formulation devised by Puleo and Milenkovic [2015] is as follows:

$$\text{LP}_{\text{CSBCC}}: \quad \underset{x,y}{\text{minimize}} \quad \sum_{e \in E} \left(w_e^+ x_e + w_e^- (1 - x_e) \right) + \sum_{u \in V} \varphi_u y_u$$

$$\text{subject to} \quad x_{uv} \leq x_{uz} + x_{zv}, \quad \forall u, v, z \in V, |\{u, v, z\}| = 3 \quad \text{(I)}$$

$$\sum_{v \in V, u \neq v} (1 - x_{uv}) \leq K + y_u, \quad \forall u \in V \quad \text{(II)} \qquad \text{(2.3)}$$

$$x_e \in [0, 1], \quad \forall e \in E \quad \text{(III)}$$

$$y_u \geq 0, \quad \forall u \in V. \quad \text{(IV)}$$

In the above linear program, LP_{CSBCC} $x_e = 1$ is interpreted as meaning "the endpoints of e lie in different clusters," while the opposite interpretation holds for $x_e = 0$, i.e., "the endpoints of e lie in the same cluster." To simplify notation, it is assumed that $x_{uu} = 0$, for all vertices $u \in V$. Constraint (I) models the fact that if two edges (u, z) and (z, v) are in the same cluster, then the edge (u, v) should belong to that cluster too.

The restriction on cluster sizes is represented by Constraint (II) together with the penalty term $\sum_{u \in V} \varphi_u y_u$ in the objective function, where every y_u variable represents the amount by which the cluster containing u exceeds the size bound.

The integer restriction of the LP_{CSBCC} program, where $x_e \in \{0, 1\}, \forall e \in E$, exactly corresponds to the formulation of CLUSTER-SIZE-BOUNDED-CORRELATION-CLUSTERING reported in Problem 2.14.

Since any actual clustering yields a feasible integer solution to LP_{CSBCC}, the optimal value of LP_{CSBCC} is a lower bound for the optimal cost of the CLUSTER-SIZE-BOUNDED-CORRELATION-CLUSTERING problem.

The idea behind Puleo and Milenkovic [2015]'s region-growing algorithm is to start with an optimal solution to LP_{CSBCC} and round the solution to produce a clustering. The main result in this regard is to prove that this rounding process only increases the cost of the solution by a constant multiplicative factor, thus rendering Puleo and Milenkovic [2015]'s one a constant-factor approximation algorithm for CLUSTER-SIZE-BOUNDED-CORRELATION-CLUSTERING.

Rounding the linear-programming solution.   Given a solution $\langle \vec{x}, \vec{y} \rangle$ to LP_{CSBCC}, where $\vec{x}$ denotes the vector of all edge costs x_e, while $\vec{y}$ denotes the vector of all vertex penalties y_u, a clustering ℓ is yielded via the procedure outlined as Algorithm 2.4. The only difference between Algorithm 2.4 and the second stage of the region-growing algorithm in Charikar et al. [2005] is the use of a parameter $\alpha \in (0, \frac{1}{2}]$ rather than the fixed value of $\frac{1}{2}$.

As shown in more detail later, parameter α plays a central role in allowing Puleo and Milenkovic [2015]'s ultimate algorithm to handle an extended range of values for the edge weights. Puleo and Milenkovic [2015] prove that the clustering produced by Algorithm 2.4 has cost at most c_α times the cost of the solution to LP_{CSBCC}, where c_α is a constant depending only on α.

As anticipated above, being the cost of the solution to LP_{CSBCC} a lower bound on the cost of the optimal solution to CLUSTER-SIZE-BOUNDED-CORRELATION-CLUSTERING, it eas-

---

Algorithm 2.4 RoundLP$_{\text{CSBCC}}$ [Puleo and Milenkovic, 2015]

Input: A instance $\langle G = (V, E, w^+, w^-), \varphi, K \rangle$ of Cluster-Size-Bounded-Correlation-Clustering, a solution $\langle \vec{x}, \vec{y} \rangle$ to LP$_{\text{CSBCC}}$ on input $\langle G, \varphi, K \rangle$, $\alpha \in (0, \frac{1}{2}]$

Output: A clustering $\ell : V \to \mathbb{N}$

1: $S \leftarrow V$
2: **while** $S \neq \emptyset$ **do**
3:      pick a vertex u from S
4:      $T \leftarrow \{v \in S \setminus \{u\} \mid x_{uv} \leq \alpha\}$
5:      $C \leftarrow \{u\}$
6:      **if** $\sum_{v \in T} x_{uv} < \alpha |T|/2$ **then**
7:          $C \leftarrow C \cup T$
8:      **end if**
9:      add cluster C to the output clustering ℓ
10:     $S \leftarrow S \setminus C$
11: **end while**

---

ily follows that Algorithm 2.4 is a c_α-approximation algorithm for Cluster-Size-Bounded-Correlation-Clustering. The main result by Puleo and Milenkovic [2015] is formally stated next.

Theorem 2.15   [Puleo and Milenkovic, 2015].   *Given an instance $I = \langle G, \varphi, K \rangle$ of Cluster-Size-Bounded-Correlation-Clustering, let $\langle \vec{x}, \vec{y} \rangle$ be a solution to LP$_{\text{CSBCC}}$ on input I, with cost equal to $cost(\vec{x}, \vec{y})$, and let $\alpha \in (0, \frac{1}{2}]$. It holds that the clustering produced by Algorithm 2.4 on input $\langle I, \vec{x}, \vec{y}, \alpha \rangle$ has cost at most $c_\alpha cost(\vec{x}, \vec{y})$, where*

$$c_\alpha = \max \left\{ \varphi^*, \frac{2\alpha\varphi^*}{1 - 2\alpha} + \frac{1}{1 - 2\alpha + \frac{\alpha}{2\tau}}, \frac{2}{\alpha} \right\},$$

and $\varphi^* = \max_{u,v \in V, u \neq v}(\varphi_u + \varphi_v)$. Also, if $\varphi^* = 0$ and $\alpha = \frac{1}{2}$, then the same result holds with $c_\alpha = \max \left\{ \frac{1}{1 - 2\alpha + \frac{\alpha}{2\tau}}, \frac{2}{\alpha} \right\}$.

The proof of Theorem 2.15 follows the outline of the analysis performed by Charikar et al. [2005] for their 4-approximation algorithm for Min-Disagree.

The time complexity of solving the linear program LP$_{\text{CSBCC}}$ and running Algorithm 2.4 is as follows. Interior-point linear-programming solvers based on Karmarkar's method [Karmarkar and Ramakrishnan, 1991, Mehrotra, 1992] requires $\mathcal{O}(N_v^{3.5}|I|^2 \log |I| \log \log |I|)$ operations in the worst case, where N_v is the number of variables, and $|I|$ is the size of the problem input. For the case of Cluster-Size-Bounded-Correlation-Clustering at hand, N_v is $\mathcal{O}(|V|^2)$. Algorithm 2.4 takes instead $\mathcal{O}(|V| + |E|)$ time.

Insights into the approximation factor c_α.   An important point about the approximation factor c_α of Algorithm 2.4 shown in Theorem 2.15 is to determine the value of the input parameter α (in terms of φ^* and τ) that minimizes c_α. This optimal value of α can be computed by solving the following:

$$\frac{2\alpha\varphi^*}{1-2\alpha} + \frac{1}{1-2\alpha+\frac{\alpha}{2\tau}} = \max\left\{\varphi^*, \frac{2}{\alpha}\right\}. \tag{2.4}$$

This is a cubic equation in α that has an unwieldy analytical solution. Hence, rather than seeking such a general analytical solution, which would be overly complicated to express and would likely have limited practical usability, Puleo and Milenkovic [2015] give a more informed idea of the approximation factor c_α by providing a number of particular parameter values for which Equation (2.4) admits a simple explicit solution.

First of all, the restriction $\varphi^* = 0$ is considered. In that case Theorem 2.15 yields a simpler approximation factor of $\max\left\{\frac{1}{1-2\alpha+\frac{\alpha}{2\tau}}, \frac{2}{\alpha}\right\}$, which leads to an optimal value of α given by the solution to

$$\frac{1}{1-2\alpha+\frac{\alpha}{2\tau}} = \frac{2}{\alpha}.$$

Hence, in this case the resulting optimal value of α equals $\frac{2\tau}{5\tau-1}$, with the corresponding approximation ratio being equal to $c_\alpha = 5 - \frac{1}{\tau}$. If, additionally, $\tau = 1$, then a slight generalization of the probability constraint is being enforced, and the resulting approximation factor becomes equal to 4, as in Charikar et al. [2005]. Indeed, in this special case (i.e., $\varphi^* = 0$, $\tau = 1$, and $\alpha = \frac{1}{2}$), Algorithm 2.4 reduces exactly to the one devised in Charikar et al. [2005].

This is the observation already reported above, i.e., that Charikar et al. [2005]'s algorithm works as is and achieves the same approximation factor for weighted graphs with weights satisfying the probability constraint. Instead, taking into account the limit $\tau \to \infty$ means allowing the weights w_e^- to be arbitrarily large. In this case, with still $\varphi^* = 0$, the optimal value of α is $\frac{2}{5}$, yielding an approximation factor 5.

When a restriction $\varphi^* \in (0, 4]$ is assumed, it holds that $\max\left\{\varphi^*, \frac{2}{\alpha}\right\} = \frac{2}{\alpha}$, as $\alpha \le \frac{1}{2}$. This way, Equation (2.4) becomes

$$\frac{2\alpha\varphi^*}{1-2\alpha} + \frac{1}{1-2\alpha+\frac{\alpha}{2\tau}} = \frac{2}{\alpha},$$

which can be shown to have, for any $\tau \ge 0$ and any $\varphi^* \in (0, 4]$, a unique solution on $\alpha \in (0, \frac{1}{2})$. In particular, for $\tau \to \infty$, the optimal α is equal to $\frac{-5+\sqrt{25+16\varphi^*}}{4\varphi^*}$.

When, additionally, $\varphi^* = 2$, it means that a hard constraint on the cluster sizes is being required: in this case the approximation factor is roughly equal to 6.275.

Finally, setting $\varphi^* = 2$ and $\tau = 1$ corresponds to enforcing both a probability constraint on the edge weights and a hard constraint on the cluster sizes. In this case the optimal choice for α is $\frac{1}{3}$, and the resulting approximation factor is 5.

Remarks.   Puleo and Milenkovic [2015] point out that their strategy in devising an approximation algorithm for CLUSTER-SIZE-BOUNDED-CORRELATION-CLUSTERING has consisted in incorporating the desired cluster-size bounds into both the linear program and the rounding procedure.

A different strategy corresponds to solving the linear program without any size-bound constraints, and imposing the size bounds only in the rounding constraints.

Preliminary work has shown that constant-factor approximation algorithms can also be devised along these lines, under the same weight regime, and that such algorithms can avoid a pathological behavior that is sometimes exhibited by Algorithm 2.4. Future work will pursue this line of inquiry further.

Another remarkable observation by Puleo and Milenkovic [2015] is a curious asymmetry in their result. In fact, Puleo and Milenkovic [2015]'s algorithm is capable of handling size-bounded clustering instances with arbitrarily large negative weights, but cannot handle instances with very large positive weights. Since clustering is hard when the weights are allowed to be arbitrary, one cannot reasonably hope to handle instances with both large negative weights and large positive weights, but it is reasonable to ask whether there is a constant-factor approximation algorithm for instances with arbitrarily large positive weights and bounded negative weight, i.e., the mirror image, in some sense, of the instances handled by Puleo and Milenkovic [2015]'s algorithm.

However, the hardness of the BALANCED-GRAPH-PARTITIONING problem[6] suggests that no such algorithm is possible: since only positive errors matter in BALANCED-GRAPH-PARTITIONING, that problem can be roughly modeled as a weighted correlation-clustering problem by giving each positive edge a weight of $W|E|$, and each negative edge weight 1, with W being a positive constant large enough to guarantee that a single positive error costs much more than the possible negative errors altogether.

Andreev and Räcke [2006] prove that it is strongly **NP**-hard to achieve any finite approximation ratio for the BALANCED-GRAPH-PARTITIONING problem when a partition into k clusters is required, with each cluster being of size exactly $\frac{n}{k}$. This suggests that it should also be **NP**-hard to solve the corresponding correlation-clustering instances with an upper bound of size $\frac{n}{k}$ for clusters, and, due to the strong **NP**-hardness, the existence of pseudo-polynomial-time algorithms is ruled out as well. However, the analogy is not perfect, since solutions to the correlation-clustering instance may use more than k clusters, therefore the question deserves further investigation.

2.2.2   A (RANDOMIZED) PIVOTING ALGORITHM

A further contribution by Puleo and Milenkovic [2015] is a randomized pivoting algorithm for CLUSTER-SIZE-BOUNDED-CORRELATION-CLUSTERING, for the special case of *unweighted*

---

[6]BALANCED-GRAPH-PARTITIONING is the problem of partitioning a graph into k components of roughly equal size, while minimizing the capacity of the edges between different components of the resulting cut [Andreev and Räcke, 2006].

graphs and *hard cluster-size bounds*. This algorithm—termed CSB-CC-pivot—is inspired by the traditional QwickCluster pivoting algorithm for MIN-DISAGREE by Ailon et al. [2008a] (Section 1.5), and is shown to achieve an (expected) approximation factor of 7. Note that the setting unweighted graphs and hard cluster-size bounds falls into the case $\varphi^* = 2$ and $\tau = 1$ discussed in Section 2.2.1. For that case, the region-growing algorithm by Puleo and Milenkovic [2015] is shown to be a 6-approximation algorithm. However, despite the pivoting algorithm is worse in terms of approximation guarantee, it is still a relevant contribution, as it does not require any linear-programming solver, thus being faster and easier-to-implement than the region-growing algorithm.

CSB-CC-pivot takes as input an unweighted (signed) graph. Adopting the general notation of Problem 2.14, this corresponds to assume that the edge set of the graph G in input to CSB-CC-pivot is partitioned into $E = E^+ \cup E^-$, such that $w_e^+ = 1, w_e^- = 0$ for all $e \in E^+$, and $w_e^+ = 0, w_e^- = 1$ for all $e \in E^-$. Moreover, as CSB-CC-pivot handles hard cluster-size bounds only, the vertex penalties are assumed to be $\varphi_u = 1$, for all $u \in V$. Puleo and Milenkovic [2015]'s CSB-CC-pivot algorithm builds upon he popular CC-pivot algorithm by Ailon et al. [2008a], in the following simple way. It first computes a subgraph G' of the input graph G by removing a minimal set $E' \subseteq E^+$ of positive edges to ensure that every vertex has a number of positive adjacent edges at most equal to the desired cluster-size bound K. Then, it simply runs CC-pivot on G', taking advantage from the fact that CC-pivot always clusters a vertex with vertices connected to it by positive edges, thus guaranteeing that the resulting clustering satisfies the cluster-size bound. The set E' of edges to be removed can be computed in $\mathcal{O}(\sqrt{K|V|}|V|^2)$ time [Hell and Kirkpatrick, 1993], which corresponds to the overall time complexity of CSB-CC-pivot, as this dominates over the $\mathcal{O}(|V| + |E|)$ time taken by CC-pivot.

The approximation guarantee of CSB-CC-pivot is as follows.

Theorem 2.16 [**Puleo and Milenkovic, 2015**].   *Let G be an unweighted signed graph, let φ be a function assigning a penalty equal to 1 to every vertex of G, and let K a nonnegative integer. CSB-CC-pivot is an expected 7–approximation algorithm for* CLUSTER-SIZE-BOUNDED-CORRELATION-CLUSTERING *on input $\langle G, \varphi, K \rangle$.*

Proof. Let OPT$_G$ and OPT$_{G'}$ denote the optimal size-bounded clustering costs in G and G', respectively. Observe that OPT$_G \geq$ OPT$_{G'} - |E'|$, since for any clustering ℓ, $cost_G(\ell) \geq cost_{G'}(\ell) - |E'|$, where $cost_{\hat{G}}(\hat{\ell})$ denotes the cost of clustering $\hat{\ell}$ on graph $\hat{G}$. Moreover, OPT$_G \geq |E'|$, as the positive edges contained within clusters constitute a subgraph of maximum degree at most K. Taking a convex combination of these lower bounds yields the following lower bound on OPT$_G$:

$$\text{OPT}_G \geq \frac{3}{7}(\text{OPT}_{G'} - |E'|) + \frac{4}{7}|E'| = \frac{1}{7}(3\text{OPT}_{G'} + |E'|).$$

On the other hand, for any clustering ℓ, it holds that $cost_G(\ell) \leq cost_{G'}(\ell) + |E'|$, and it is known (by Ailon et al. [2008a]) that $\mathbb{E}[cost_{G'}(\ell_{ccp})] \leq 3\text{OPT}_{G'}$, where ℓ_{ccp} is the clustering yielded by the CC-pivot algorithm. Hence,

$$\mathbb{E}[cost_G(\ell_{ccp})] \ \leq \ \mathbb{E}[cost_{G'}(\ell_{ccp})] + |E'| \ \leq \ 3\mathsf{OPT}_{G'} + |E'| \ \leq \ 7\mathsf{OPT}_G.$$

$\square$

2.2.3   NON-UNIFORM CONSTRAINED CLUSTER SIZES

In Puleo and Milenkovic [2015]'s problem all the clusters share the same upper bound on their cluster size and that upper bound can be violated by paying the corresponding penalty cost.

Ji et al. [2020] introduce a variant of Puleo and Milenkovic [2015]'s problem where not all the upper bounds are the same and not all such upper bounds can be violated. Specifically, Ji et al. [2020] define the problem where, given a positive integer U_v for every input vertex v, the goal is to partition the vertices according to the traditional MIN-DISAGREE formulation, while also requiring for each cluster C to satisfy $|C| \leq \min_{v \in C} U_v$.

Given a parameter $\alpha \in (0, \frac{1}{2}]$, Ji et al. [2020] devise a $\left(\frac{1}{1-\alpha}, \frac{2}{\alpha}\right)$-bicriteria approximation algorithm, meaning that the solution returned by the algorithm has a cost that is at most $\frac{2}{\alpha}$ times the optimum, and for each cluster C in the solution, it is ensured that $|C| \leq \frac{1}{1-\alpha} \min_{v \in C} U_v$. Ji et al. [2020] also devise a further algorithm achieving a $2U$ approximation, where $U = \max_{v \in V} U_v$.

2.3   CORRELATION CLUSTERING WITH ERROR BOUNDS

Geerts and Ndindi [2016] study the ERROR-BOUNDED-CORRELATION-CLUSTERING problem, that is correlation clustering in presence of user-specified *error bounds* on the quality of the output clustering. Specifically, Geerts and Ndindi [2016] consider *general, edge-weighted graphs* as input to their problem.

Given a graph G of this kind, and a clustering ℓ of the vertices of G, let the *false negatives* in ℓ correspond to the positive edges of G connecting vertices that cross different clusters of ℓ. Similarly, let the *false positives* in ℓ be the negative edges of G connecting vertices within the same cluster of ℓ. Geerts and Ndindi [2016] define ERROR-BOUNDED-CORRELATION-CLUSTERING as the problem of finding a minimal set of edges of G such that, deleting them from G, a clustering of the updated G exists whose sum of weights of false negatives and false positives is no more than two user-specified thresholds μ_{fn} and μ_{fp}, respectively.

An equivalent formulation of ERROR-BOUNDED-CORRELATION-CLUSTERING, which is closer to the standard formulation of (the MIN-DISAGREE variant of) correlation clustering is as follows.

Problem 2.17   (Error-Bounded-Correlation-Clustering)   Let $G = (V,$ $E, w, \mathcal{L})$ be a weighted signed graph, where V is a set of n vertices, $E \subseteq \binom{V}{2}$

is a set of m edges, $w : E \to \mathbb{R}^+$ is a function assigning a positive weight to each edge, and $\mathcal{L} : E \to \{-1, +1\}$ is a function labeling an edge as either positive or negative. Given a clustering $\ell : V \to \mathbb{N}$ of V, let $w_{\text{fn}}(\ell)$ and $w_{\text{fp}}(\ell)$ denote the sum of the weights of false positives and false negatives in ℓ, respectively:

$$w_{\text{fn}}(\ell) = \sum_{\substack{(u,v)\in E, \\ \ell(u)=\ell(v)}} \frac{1 - \mathcal{L}(u, v)}{2} w(u, v), \qquad w_{\text{fp}}(\ell) = \sum_{\substack{(u,v)\in E, \\ \ell(u)\neq\ell(v)}} \frac{1 + \mathcal{L}(u, v)}{2} w(u, v).$$

Given a weighted signed graph $G = (V, E, w, \mathcal{L})$, and two real numbers $\mu_{\text{fn}}, \mu_{\text{fp}} \geq 0$, the objective of ERROR-BOUNDED-CORRELATION-CLUSTERING is to find a clustering ℓ that minimizes the following:

$$w(\ell) = \max\{0, w_{\text{fn}}(\ell) - \mu_{\text{fn}}\} + \max\{0, w_{\text{fp}}(\ell) - \mu_{\text{fp}}\}. \qquad (2.5)$$

The above formulation highlights that ERROR-BOUNDED-CORRELATION-CLUSTERING assumes that the two ingredients of the correlation-clustering cost may have unequal importance within the overall objective function, with this importance being established by two user-defined thresholds, i.e., μ_{fn} and μ_{fp}. Therefore, ERROR-BOUNDED-CORRELATION-CLUSTERING is a generalization of (the min-disagree formulation of) standard correlation clustering: in fact, for $\mu_{\text{fn}} = \mu_{\text{fp}} = 0$, ERROR-BOUNDED-CORRELATION-CLUSTERING boils down to the WEIGHTED-MIN-DISAGREE problem. Based on this, the following hardness result can be straightforwardly observed.

Theorem 2.18 [Geerts and Ndindi, 2016]. ERROR-BOUNDED-CORRELATION-CLUSTERING *is **NP**-hard for both weighted and unweighted graphs.*

2.3.1 CONNECTION WITH BOUNDED-MIN-MULTICUT

The main result by Geerts and Ndindi [2016] is an approximation algorithm based on the well-established region-growing technique. To devise such an algorithm, a connection with a variant of the MIN-MULTICUT problem—termed BOUNDED-MIN-MULTICUT—is firstly shown.

The standard MIN-MULTICUT problem is as follows: given an edge-weighted graph $G = (V, E, w)$, and a set $S = \{(s_i, t_i) \mid s_i, t_i \in V\}_{i=1}^k$ of k source-sink vertex pairs, find a subset $T \subseteq E$ of edges (i.e., a so-called *multicut*) such that (i) the removal of the edges in T from G disconnects all pairs in S, and (ii) the sum $w(T)$ of the weights on the edges in T is minimum [Hu, 1963]. The BOUNDED-MIN-MULTICUT problem differs from MIN-MULTICUT in that it works on signed graphs, and bounds (i.e., real numbers) $\mu^+, \mu^- \geq 0$ are present that limit the allowed positive and negative edges, respectively, in a multicut. Denoting by T^+ and T^- the set of positive and negative edges within a multicut T, respectively, T is said *valid* if $w(T^+) \leq \mu^+$ and $w(T^-) \leq \mu^-$. A valid multicut may not always exist. Hence, BOUNDED-MIN-MULTICUT asks for a "minimal" set of edges (i.e., a set of edges whose sum of weights in minimum) to be deleted such that the existence of a valid multicut is guaranteed.

Geerts and Ndindi [2016] show a correspondence between ERROR-BOUNDED-CORRELATION-CLUSTERING and BOUNDED-MIN-MULTICUT in a way similar to the correspondence between MIN-MULTICUT and MIN-DISAGREE provided by Demaine et al. [2006]. In particular, given an instance $\langle G = (V, E, w, \mathcal{L}), \mu_{\mathrm{fn}}, \mu_{\mathrm{fp}} \rangle$ of ERROR-BOUNDED-CORRELATION-CLUSTERING, the corresponding, equivalent instance $\langle G' = (V', E', w', \mathcal{L}'), S, \mu^+, \mu^- \rangle$ of BOUNDED-MIN-MULTICUT is obtained as outlined in Algorithm 2.5.

Algorithm 2.5 BCC2BMC [Geerts and Ndindi, 2016]

Input: An instance $\langle G = (V, E, w, \mathcal{L}), \mu_{\mathrm{fn}}, \mu_{\mathrm{fp}} \rangle$ of ERROR-BOUNDED-CORRELATION-CLUSTERING

Output: An instance $\langle G' = (V', E', w', \mathcal{L}'), S, \mu^+, \mu^- \rangle$ of BOUNDED-MIN-MULTICUT

1: For every negative edge (u, v), introduce a new vertex $z_{u,v}$. Set V' to be equal to V together with these newly added vertices

2: Let the positive edges in E' correspond to all and only the positive edges in E, and let their weight be equal to the weight of the corresponding original edge in E

3: Let the negative edges in E' consist of one new negative edge $(z_{u,v}, u)$ for each vertex of the form $z_{u,v}$, and set the weight $w'(z_{u,v}, u)$ of every $(z_{u,v}, u)$ equal to $w(u, v)$

4: Let S consist of the source-sink pairs $(z_{u,v}, v)$, for every newly added vertex $z_{u,v}$

5: Set $\mu^+ = \mu_{\mathrm{fn}}$ and $\mu^- = \mu_{\mathrm{fp}}$

A remarkable observation about the transformation in Algorithm 2.5 is that it differs from the one between MIN-MULTICUT and MIN-DISAGREE devised in Demaine et al. [2006] only in the fact that the $z_{u,v}$ edges are marked as negatives, while the remaining edges are marked as positive.

The following main result is shown in Geerts and Ndindi [2016].

Theorem 2.19 **[Geerts and Ndindi, 2016].** *Given an instance $I = \langle G = (V, E, w, \mathcal{L}), \mu_{\mathrm{fn}}, \mu_{\mathrm{fp}} \rangle$ of* ERROR-BOUNDED-CORRELATION-CLUSTERING, *and the corresponding instance $I' = \langle G' = (V', E', w', \mathcal{L}'), S, \mu^+, \mu^- \rangle$ of* BOUNDED-MIN-MULTICUT *constructed as in Algorithm 2.5, a cost-0 solution to* ERROR-BOUNDED-CORRELATION-CLUSTERING *on input I corresponds to a valid multicut on input I', and vice versa.*

Proof. The proof is analogous to Lemmas 4.5–4.6 in Demaine et al. [2006]. Let ℓ be a cost-0 solution to ERROR-BOUNDED-CORRELATION-CLUSTERING on input I, and let a multicut T

on input I' consist of all positive edges in E' that contribute to $w_{\text{fn}}(\ell)$, and all negative edges $(z_{u,v}, u) \in E'$ corresponding to negative edges in E that contribute to $w_{\text{fp}}(\ell)$. It is shown in Demaine et al. [2006] that T is a multicut. Moreover, observe that if ℓ has cost 0, then the sum of the weights on the positive edges in T is guaranteed to be $\leq \mu^+$, and the sum of the weights on the negative edges in T is guaranteed to be $\leq \mu^-$. Hence, T is a valid multicut for the instance I'.

Similarly, given a multicut T that is valid for I', one can construct a clustering ℓ of G as follows. Let T' be the union of the set of positive edges in T and the set of negative edges $(u, v) \in E$ corresponding to a negative edge $(z_{u,v}, u) \in T$. Denote by G_T^+ the subgraph of G induced by all positive edges of G that do not belong to T. Then, ℓ is defined as the set of all connected components of G_T^+. Demaine et al. [2006] show that T' consists of the false positives and false negatives of the clustering ℓ (on input I). Therefore, $w_{\text{fp}}(\ell) \leq \mu_{\text{fp}}$ and $w_{\text{fn}}(\ell) \leq \mu_{\text{fn}}$, meaning that ℓ is a cost-0 solution to Error-Bounded-Correlation-Clustering on input I. $\qquad\square$

Based on the above theorem, Geerts and Ndindi [2016] further observe that for a solution ℓ to Error-Bounded-Correlation-Clustering on input I there exists a solution T to Bounded-Min-Multicut on input I' (where I' is constructed from I as in Algorithm 2.5) such that $w(\ell) = w'(T)$, and vice versa. With Theorem 2.19 and this observation in place, the following ultimate result becomes immediate: any (approximation) algorithm for Bounded-Min-Multicut is an (approximation) algorithm for Error-Bounded-Correlation-Clustering. This result provides the basis for the design of an approximation algorithm for Error-Bounded-Correlation-Clustering, whose details are reported next.

2.3.2   AN APPROXIMATION ALGORITHM

The approximation algorithm for Error-Bounded-Correlation-Clustering devised by Geerts and Ndindi [2016] is based on a relaxation of an integer-programming formulation of Bounded-Min-Multicut.

Given an edge-weighted signed graph $G = (V, E, w, \mathcal{L})$, a set $S = \{(s_i, t_i) \mid s_i, t_i \in V\}_{i=1}^k$ of k source-sink vertex pairs, and two real numbers $\mu^+, \mu^- \geq 0$, the integer program for Error-Bounded-Correlation-Clustering, denoted by IP_{BMC}, is as follows:

$$
\begin{aligned}
\text{IP}_{\text{BMC}}: \quad & \underset{x,y}{\text{minimize}} \quad \sum_{(u,v)\in E} w_{uv}(x_{uv} - y_{uv}) \\
& \text{subject to} \quad \sum_{(u,v)\in P_i} x_{uv} \geq 1, \ \forall P_i \in \mathcal{P}_i, 1 \leq i \leq k \quad \text{(I)} \\
& \qquad\qquad \sum_{(u,v)\in E^+} w_{uv}\, y_{uv} \leq \mu^+ \quad \text{(II)} \\
& \qquad\qquad \sum_{(u,v)\in E^-} w_{uv}\, y_{uv} \leq \mu^- \quad \text{(III)} \\
& \qquad\qquad x_{uv} \geq y_{uv}, \ \forall(u,v) \in E \quad \text{(IV)} \\
& \qquad\qquad x_{uv}, y_{uv} \in \{0,1\}, \ \forall(u,v) \in E, \quad \text{(V)}
\end{aligned}
\qquad (2.6)
$$

where $\mathcal{P}_i$ denotes the set of all paths from s_i to t_i in G, for all $i \in [1, k]$, $E^+ \subseteq E$ and $E^- \subseteq E$ are the set of positive and negative edges of G, respectively, and w_{uv} is a short form of $w(u, v)$.

The IP_{BMC} program is a simple modification of the standard integer program for MIN-MULTICUT [Vazirani, 2001]. Specifically, the integer program IP_{MC} for MIN-MULTICUT can be obtained from IP_{BMC} by setting $\mu^+ = \mu^- = 0$, i.e., by ignoring the y_{uv} variables. Moreover, note that IP_{BMC} has an exponential number of constraints, but, similarly as in the MIN-MULTICUT case, it can be converted into an equivalent one with polynomial size, as described in Garg et al. [1996]. Such a conversion simply consists in introducing further binary variables z_u^i, one for each vertex $u \in V$ and each $(s_i, t_i) \in S$, and replacing Constraint (I) in IP_{BMC} with the following two ones:

$$z_u^i - z_v^i \leq x_{uv}, \ \forall (u, v) \in E, 1 \leq i \leq k \quad (\text{I}')$$
$$z_{s_i}^i - z_{t_i}^i \geq 1, \ \forall (s_i, t_i) \in S \quad (\text{I}'').$$

To see the equivalence between Constraint (I) and Constraints (I')—(I''), consider a source-sink pair $(s_i, t_i) \in S$, and assume that there is a path P of size $|P| = p + 1$ between s_i and t_i consisting of edges $(s_i, v_1), (v_1, v_2), \ldots, (v_p, t_i)$. Based on Constraint (I'), it holds that

$$1 \leq z_{s_i}^i - z_{t_i}^i = z_{s_i}^i - z_{v_1}^i + z_{v_1}^i - z_{v_2}^i + \cdots - z_{v_p}^i + z_{v_p}^i - z_{t_i}^i,$$

while Constraint (I'') states that

$$z_{s_i}^i - z_{v_1}^i \leq x_{s_i v_1}, \ z_{v_1}^i - z_{v_2}^i \leq x_{v_1 v_2}, \ \ldots, \ z_{v_p}^i - x_{t_i}^i \leq x_{v_p t_i}.$$

Hence,

$$1 \leq z_{s_i}^i - z_{t_i}^i \leq \sum_{(u,v) \in P} x_{uv}.$$

This holds for every path P between every source-sink pair. Therefore, Constraint (I) is satisfied. As far as the converse, assume that Constraint (I) is satisfied. For each $(s_i, t_i) \in S$, set $z_u^i = \sum_{(u',v') \in P_{ut_i}^*} x_{u'v'}$, where $P_{ut_i}^*$ is the shortest path between u and t_i. This implies that $z_{s_i}^i \geq 1$ by Constraint (I), and $z_{t_i}^i = 0$, as t_i lies at distance 0 from itself, which satisfies Constraint (I'). Also, note that for an edge $(u, v) \in E$ it holds that $z_u^i - z_v^i = x_{uv}$, which is in accordance with Constraint (I'').

The BMulticut Algorithm

Geerts and Ndindi [2016] exploits the above integer-programming formulation IP_{BMC} to devise an approximation algorithm for BOUNDED-MIN-MULTICUT that is inspired by the region-growing algorithm for MIN-MULTICUT due to Garg et al. [1996]. That region-growing algorithm solves the relaxation of the aforementioned integer program IP_{MC} for MIN-MULTICUT, and repeatedly grows regions until all source-sink pairs have become disconnected. The edges adjacent to the identified regions constitute the ultimate output multicut. Similarly, Geerts and Ndindi [2016]'s algorithm, termed BMulticut, solves the relaxation of the IP_{BMC}

Algorithm 2.6 BMulticut [Geerts and Ndindi, 2016]

Input: An edge-weighted signed graph $G = (V, E, w, \mathcal{L})$, a set $S = \{(s_i, t_i)\}_{i=1}^{k}$ of vertex pairs, $\mu^+, \mu^- \in \mathbb{R}$, $R \in \mathbb{N}$

Output: $\Delta E \subseteq E$

1: $\vec{d}_0 \leftarrow$ SolveLP(G, S, μ^+, μ^-)
2: $\langle G_1, S_1, B_0 \rangle \leftarrow$ GrowRegion$(G, S, \vec{d}_0, |S|, 1)$ // Algorithm 2.7
3: $\partial B_0 \leftarrow \{e \in E \mid |e \cap B_0| = 1\}$
4: $\mu_1^+ \leftarrow \max\{0, \mu^+ - w(\partial B_0 \cap E^+)\}$, $\mu_1^- \leftarrow \max\{0, \mu^- - w(\partial B_0 \cap E^-)\}$
5: $i \leftarrow 1$, $next \leftarrow$ true
6: **while** $(\mu_i^+ \neq \mu_{i-1}^+ \vee \mu_i^- \neq \mu_{i-1}^-) \wedge |S_i| > 0 \wedge i \leq R \wedge next$ **do**
7:     $\vec{d}_i \leftarrow$ SolveLP$(G_i, S_i, \mu_i^+, \mu_i^-)$
8:     **if** $\sum_{e \in E_i} \vec{d}_i[e] \leq \sum_{e \in E_i} \vec{d}_{i-1}[e]$ **then**
9:       $\langle G_{i+1}, S_{i+1}, B_i \rangle \leftarrow$ GrowRegion$(G_i, S_i, \vec{d}_i, |S|, 1)$ // Algorithm 2.7
10:       $\partial B_i \leftarrow \{e \in E_i \mid |e \cap B_i| = 1\}$
11:       $\mu_{i+1}^+ \leftarrow \max\{0, \mu_i^+ - w(\partial B_i \cap E_i^+)\}$, $\mu_{i+1}^- \leftarrow \max\{0, \mu_i^- - w(\partial B_i \cap E_i^-)\}$
12:       $i \leftarrow i + 1$
13:     **else**
14:       $next \leftarrow$ false
15:     **end if**
16: **end while**
17: **for** $j = 0, \ldots, i-1$ **do**
18:     $\langle H, S, T_j \rangle \leftarrow$ GrowRegion$(G_{j+1}, S_{j+1}, \vec{d}_j, |S|, +\infty)$ // Algorithm 2.7
19:     $cut_j \leftarrow \partial B_0 \cup \cdots \cup \partial B_j \cup T_j$
20:     $\Delta E_j \leftarrow$ ExtractDeltaE(cut_j)
21: **end for**
22: $j^* \leftarrow \arg\min_{j \in [0, i-1]} w(\Delta E_j)$
23: $\Delta E \leftarrow \Delta E_{j^*}$

---

integer program, but, differently from Garg et al. [1996]'s algorithm, it also employs an adaptive optimization strategy in which the underlying integer program (and its relaxation) is updated between two consecutive region-growing steps, provided that this is expected to lead to a better solution. Thus, while Garg et al. [1996]'s algorithm solves the relaxed integer program only once (before the region-growing process starts), the BMulticut algorithm solves its relaxed integer program multiple times. Since solving linear programs comes at a cost, the number of linear programs to be solved is limited by a parameter K. Solutions to the BOUNDED-MIN-MULTICUT problem are then obtained from the produced multicut T by a post-processing step. That is, T is split into ΔE such that $T \setminus \Delta E$ is a valid multicut with respect to the given bounds μ^+ and μ^-.

Algorithm 2.7 GrowRegion [Geerts and Ndindi, 2016]

Input: An edge-weighted signed graph $G = (V, E, w, \mathcal{L})$, a set $S = \{(s_i, t_i)\}_{i=1}^k$ of vertex pairs, an $|E|$-dimensional real-valued vector $\vec{d}$ (with $\vec{d}[e] \in [0, 1], \forall e \in E$), $h \in \mathbb{N}, N \in \mathbb{R}$
Output: A subgraph H of G, $S' \subseteq S$, $T \subseteq E$

1: $F \leftarrow \sum_{e \in E} w_e \vec{d}[e]$
2: $\varepsilon \leftarrow 2\ln(h + 1)$
3: $H \leftarrow G, \quad T \leftarrow \emptyset, \quad i = 1, \quad S' \leftarrow S$
4: **while** $|S'| > 0 \wedge i \leq N$ **do**
5: $\quad$ $grow \leftarrow$ true
6: $\quad$ pick a source-sink pair (s, t) from S' and let $region = \emptyset$
7: $\quad$ let L be a list containing the vertices of H, sorted by their increasing $\vec{d}$-distance to s; assume that s is the first element $L[0]$ in this list and let $L = L[0]$
8: $\quad$ **while** $grow$ **do**
9: $\quad\quad$ $region \leftarrow region \cup L$
10: $\quad\quad$ $\mathcal{V}(region) \leftarrow F/h + \sum_{e \in E : e \cap region \neq \emptyset} w_e \vec{d}[e]$
11: $\quad\quad$ $c(region) \leftarrow \sum_{e \in E : |e \cap region| = 1} w_e$
12: $\quad\quad$ **if** $c(region) \leq \varepsilon \mathcal{V}(region)$ **then**
13: $\quad\quad\quad$ $grow \leftarrow$ false
14: $\quad\quad$ **else**
15: $\quad\quad\quad$ $L \leftarrow L.next$
16: $\quad\quad$ **end if**
17: $\quad$ **end while**
18: $\quad$ $H \leftarrow$ remove from H all vertices (and incident edges) in $region$
19: $\quad$ $S' \leftarrow$ remove from S' all pairs that are disconnected in H
20: $\quad$ $T \leftarrow T \cup \{e \in E \mid |e \cap region| = 1\}$
21: $\quad$ $i \leftarrow i + 1$
22: **end while**

The set ΔE is ultimately returned as output. Geerts and Ndindi [2016] note that, due to the nature of the region-growing process, it does not necessarily hold that increasing K leads to better solutions. For this reason, for a given R, the BMulticut algorithm runs (at most) R region-growing processes, corresponding to the parameter values $1, 2, \ldots, R$, and takes the best solution among the ones yielded by these processes. This clearly guarantees that the quality of solutions never degrades with increasing R.

The pseudocodes of the BMulticut algorithm and its main GrowRegion subroutine are reported in Algorithms 2.6 and 2.7, respectively. BMulticut makes also use of the following other subroutines:

- SolveLP, which solves a relaxation of the $\mathrm{IP}_{\mathrm{BMC}}$ integer program obtained by replacing Constraint (V) in $\mathrm{IP}_{\mathrm{BMC}}$ with $x_{uv}, y_{uv} \in [0, 1], \forall (u, v) \in E$. This subroutine returns an $|E|$-dimensional real-valued vector $\vec{d}$, whose entries $\vec{d}[e]$ correspond to $x_{uv}, \forall e = (u, v) \in E$.

- ExtractDeltaE, which extracts from an input subset cut_j of edges a minimal set ΔE_j of edges such that the remaining edges in cut_j form a valid multicut with respect to parameters μ^+ and μ^-. This subroutine simply removes edges from cut_j until $w(cut_j \cap E^+) \le \mu^+$ and $w(cut_j \cap E^-) \le \mu^-$ hold. The removed edges are ultimately returned as ΔE_j.

Due to the aforementioned equivalence between ERROR-BOUNDED-CORRELATION-CLUSTERING and BOUNDED-MIN-MULTICUT, the ultimate algorithm for ERROR-BOUNDED-CORRELATION-CLUSTERING can be obtained by simply (i) transforming the input ERROR-BOUNDED-CORRELATION-CLUSTERING instance into an equivalent BOUNDED-MIN-MULTICUT instance as stated in Algorithm 2.5, (ii) running BMulticut on the latter, and (iii) performing the following simple complementary steps to compute and output a clustering of the input graph (out of the various intermediate ΔE_j that have been yielded during the execution of the BMulticut algorithm):

- $\ell_j \leftarrow$ ComputeClustering(G, cut_j) (to be added right after Line 20 of Algorithm 2.6); and

- ultimately output clustering ℓ_{j*} (to be added at the very end of Algorithm 2.6).

The BMulticut algorithm is shown to achieve the following approximation guarantee.

Theorem 2.20 [**Geerts and Ndindi, 2016**]. *BMulticut is an $\mathcal{O}(\log |E^-|)$-approximation algorithm for* ERROR-BOUNDED-CORRELATION-CLUSTERING *on input* $\langle G = (V, E, w, \mathcal{L}), \mu_{\mathrm{fp}}, \mu_{\mathrm{fn}}\rangle$, *where* $E^- = \{e \in E \mid \mathcal{L}(e) = -1\}$.

The proof of the above theorem is based on showing that BMulticut is an $\mathcal{O}(\log |S|)$-approximation algorithm for BOUNDED-MIN-MULTICUT on the input instance $\langle G' = (V', E', w', \mathcal{L}'), S, \mu^+, \mu^-\rangle$ derived by the procedure for converting ERROR-BOUNDED-CORRELATION-CLUSTERING into BOUNDED-MIN-MULTICUT shown in Algorithm 2.5. This finding follows from a simple modification of the approximation guarantee of Garg et al. [1996]'s algorithm for MIN-MULTICUT. With this result in place, the ultimate $\mathcal{O}(\log |E^-|)$-approximation for ERROR-BOUNDED-CORRELATION-CLUSTERING easily derives from the fact that, according to Algorithm 2.5, S corresponds to the negative edges in E.

CHAPTER 3

Relaxed Formulations

This chapter deals with *relaxed* formulations of correlation clustering, i.e., formulations where some constraints of the basic formulation are discarded or required in a less restrictive form. This results in optimization problems whose feasible region is a superset of one of the basic correlation-clustering formulations.

In particular, this chapter discusses correlation clustering where the output clusters are allowed to overlap (Section 3.1), correlation clustering where (dis)agreements are considered locally, at the level of a single vertex, and the goal is to optimize an aggregation function of the local (dis)agreements (Section 3.2), and correlation clustering with outliers (Section 3.3).

3.1   OVERLAPPING CORRELATION CLUSTERING

In many real-world applications, it is desirable to allow overlapping clusters as objects may intrinsically belong to more than one cluster. For example, in social networks users belong to numerous communities. In biology, a large fraction of proteins belongs to several protein complexes simultaneously, and genes have multiple coding functions and participate in different metabolic pathways. In information retrieval and text mining, documents, news articles, and web pages can belong to different categories.

The problem of *Overlapping Correlation Clustering* was introduced by Bonchi et al. [2011, 2013b] as the problem of mapping each input object to a small set of labels that represent cluster membership. The number of labels does not have to be the same for all objects. The objective is to find a mapping so that the similarity between any pair of objects in the input dataset agrees as much as possible with the similarity of their corresponding sets of labels. More formally, consider a set of n objects $V = \{v_1, \ldots, v_n\}$ and a similarity value $s(u, v)$ for each pair $(u, v) \in V \times V$ of objects. In the most general case, the similarity function s takes values in the interval $[0, 1]$, but Bonchi et al. [2011, 2013b] also study the special case in which the similarity function takes only values in the set $\{0, 1\}$.

The main idea to extend the (min-disagreement) formulation of correlation clustering in order to allow overlapping of clusters is to redefine a mapping function ℓ. Instead of mapping each object to a single cluster label $c \in L$, the function ℓ is relaxed so that it can map objects to any subset of cluster labels. Thus, denoting by $\mathcal{C}_+$ the collection of all subsets of L except the empty set, i.e., $\mathcal{C}_+ = 2^L \setminus \{\emptyset\}$, one can now define the multi-labeling function ℓ to be $\ell : V \to \mathcal{C}_+$. If an object v is mapped under ℓ to a set of cluster labels $\ell(v) = \{c_1, \ldots, c_s\} \in \mathcal{C}_+$, then it is said that v participates in all the clusters $c_1, \ldots, c_s$.

For a good clustering, similar objects should be mapped to similar sets of cluster labels. Thus, to evaluate a solution to overlapping clustering, a similarity function $H : \mathcal{C}_+ \times \mathcal{C}_+ \to [0, 1]$ between sets of cluster labels is introduced. We now have the necessary ingredients to formally define the problem of overlapping clustering.

Problem 3.1   (Overlapping-Correlation-Clustering)   Given n objects $V = \{v_1, \ldots, v_n\}$, a pairwise similarity function s over $V \times V$, and a similarity function H between sets, find a multi-labeling function $\ell : V \to \mathcal{C}_+$ that minimizes the cost

$$C_{\mathrm{occ}}(V, \ell) = \sum_{(u,v) \in V \times V} |H(\ell(u), \ell(v)) - s(u, v)|. \qquad (3.1)$$

This definition of clustering aims at finding a multi-labeling that, to the highest possible degree, maintains the similarities between objects. Note that considering the error term $|H - s|$ is meaningful since both H and s are similarity functions that take values in the range $[0, 1]$. To make the problem concrete, we need to instantiate the similarity function H between two sets E, F of cluster labels. Bonchi et al. [2011, 2013b] consider two such functions: the Jaccard coefficient $J(E, F) = \frac{|E \cap F|}{|E \cup F|}$ and the set-intersection indicator function I.

$$I(E, F) = \begin{cases} 1 & \text{if } E \cap F \neq \emptyset \\ 0 & \text{otherwise.} \end{cases}$$

The Jaccard coefficient is a very natural function to measure similarity between sets and it has been used in a wide range of applications. On the other hand, in certain applications, one might be interested in whether the cluster-label sets intersect or not: for these the set-intersection indicator can be used.

3.1.1   CONSTRAINTS AND CHARACTERIZATION

So far, we have assumed a finite alphabet of labels and hence a maximum number of clusters $|L| = k$. However, for many applications, although we may have in our disposal a large number of clusters, we may not want to assign an object to all those clusters. For example, when clustering the users of a social network, we may want to use hundreds or even thousands of clusters, however, we may want to assign each user only in a handful of clusters.

Thus, Bonchi et al. [2011, 2013b] consider a second type of constraint in which we require that each object v should be mapped to at most p clusters, that is, $|\ell(v)| \leq p$ for all $v \in V$.

Therefore, when defining an instance of Problem 3.1 we have the following options:

- the similarity function s between objects may take values in the range $[0, 1]$ or it may take binary values in $\{0, 1\}$;

- the similarity function H between sets of cluster labels may be the Jaccard coefficient J or the intersection indicator I; and

- we may impose the local constraint of having at most p cluster labels for each object or we may not.

Any combination of the above options gives rise to a valid problem instance. In the following, we systematically refer to any of these problems using the notation (r, H, p), where: $r \in \{\mathtt{f}, \mathtt{b}\}$ refers to the range of the function s: $\mathtt{f}$ for fractional values and $\mathtt{b}$ binary; $H \in \{J, I\}$ refers the similarity function H: J for Jaccard coefficient and I for set-intersection; and p refers to the value of p of the local constraint, so $p = k$ means that there is no local constraint. As an example, by $(\mathtt{b}, H, k)$ we refer to two different problem instances, where s takes binary values, H can be either J or I, and there is no local constraint.

3.1.2 HARDNESS RESULTS

This section presents some hardness results derived in Bonchi et al. [2011, 2013b]. The first observation is that all instances specified by $(r, H, 1)$ correspond to the CORRELATION-CLUSTERING problem defined in Problem 1.1. The reason is that when $|\ell(v)| = 1$ for all v in V, then both the Jaccard coefficient and the intersection indicator take just 0 or 1 values. In particular, $|H(\ell(u), \ell(v)) - s(u, v)|$ becomes $1 - s(u, v)$ when $\ell(u) = \ell(v)$ and $s(u, v)$ when $\ell(u) \neq \ell(v)$. A direct consequence is that Problem 3.1 is a generalization of the standard CORRELATION-CLUSTERING problem. Since CORRELATION-CLUSTERING is **NP**-hard, and since $p = 1$ is a special case of any (r, H, p) problem, the previous observation implies that the general OVERLAPPING-CORRELATION-CLUSTERING problem is also **NP**-hard. However, in order to show that the complexity of Problem 3.1 does not derive exclusively from the hardness of the special case $p = 1$, a more general **NP**-hardness result that does not rely on such a special case, is provided next.

Theorem 3.2   [Bonchi et al., 2011, 2013b].   *The problem instances (r, I, p), with $p > 1$, are **NP**-hard.*

Proof. The idea is to show that the $(\mathtt{b}, I, p)$ problem is **NP**-hard, which also gives the **NP**-hardness for the $(\mathtt{f}, I, p)$ problem. The reduction is obtained from the problem of COVERING-BY-CLIQUES [Garey and Johnson, 1979, GT17], which is as follows: given an undirected graph $G = (V, E)$ and an integer $C \leq |E|$, decide whether G can be represented as the union of $c \leq C$ cliques. It can be shown that a zero-cost solution to the $(\mathtt{b}, I, C)$ problem identifies graphs having a covering by at most C cliques, and solutions with a cost larger than zero identify graphs that do not admit a covering by at most C cliques. To this end, given an undirected graph

$G = (V, E)$, construct an instance of the (b, I, C) problem by simply setting the set of objects to correspond to the set of vertices V. For each edge $(u, v) \in E$ set $s(u, v) = 1$, while if $(u, v) \notin E$ we set $s(u, v) = 0$. Also, set the total number of cluster labels $k = C$. At this point, the claim is easy to verify: a zero-cost solution of (b, I, C) on input (V, s) corresponds to a covering of G by at most C cliques.   □

In addition, due to the inapproximability of the COVERING-BY-CLIQUES problem, we can deduce that the problem instances (r, I, p) do not admit polynomial-time constant-factor approximation algorithms, unless **P = NP**.

The next result focuses on the problems (r, I, k), that is, using set-intersection and no local constraint, in the case where we are allowed to use a very large number of cluster labels, in particular $k = \Theta(n^2)$.

Proposition 3.3 [Bonchi et al., 2011, 2013b]   *For the problem instances $(r, I, \Theta(n^2))$, the optimal solution can be found in polynomial time.*

Proof. We start by giving each object a unique cluster label. Then we process each pair of objects for which $s(u, v) \geq \frac{1}{2}$. For any such pair of objects we make a new label, which we assign to both objects, and never use again. Thus, for pairs with $s(u, v) \geq \frac{1}{2}$, the intersection of $\ell(u)$ and $\ell(v)$ is not empty, and thus we pay $|1 - s(u, v)| \leq \frac{1}{2}$. On the other hand, for the pairs with $s(u, v) \leq \frac{1}{2}$, the intersection of $\ell(u)$ and $\ell(v)$ is empty, and thus we pay $|s(u, v)| \leq \frac{1}{2}$. Since I takes only 0/1 values, no other solution can cost less, and thus the previous process gives an optimal solution. □

When we have binary similarities, the above process straightforwardly provides a zero-cost solution.

Corollary 3.4 [Bonchi et al., 2011, 2013b]   *The problem instance $(b, I, \Theta(n^2))$ always admits a zero-cost solution that can be found in polynomial time.*

3.1.3   CONNECTION WITH GRAPH COLORING

Given that the problem instance (b, I, k) always admits a zero-cost solution, if we allow enough cluster labels, we next ask which is the minimum number of cluster labels k needed for a zero-cost solution. Bonchi et al. [2011] characterize such a number by pointing out a connection with the GRAPH-COLORING problem, whose formulation we recall next. A proper coloring of a graph $G = (V, E)$ is a function $c : V \rightarrow \{1, \ldots, k\}$ so that, for all $(u, v) \in E$, we have $c(u) \neq c(v)$. The GRAPH-COLORING problem asks to find the smallest number k—known as the *chromatic number* $\chi(G)$ of G—for which a proper coloring of G exists.

Going back to the binary (b, I, k) instance of the OVERLAPPING-CORRELATION-CLUSTERING problem, given the set of objects V and similarity function s, we consider *similar pairs* $P^+ = \{(u,v) \in V \times V \mid s(u,v) = 1\}$ and *dissimilar pairs* $P^- = \{(u,v) \in V \times V \mid s(u,v) = 0\}$. Using these, we define the graph $\widehat{G} = (P^+, \widehat{E})$, with similar pairs as vertices, and the set of edges $\widehat{E}$ given by the dissimilar pairs as follows:

$$\widehat{E} = \{((u,v),(x,y)) \in P^+ \times P^+ \mid \{(u,x),(u,y),(v,x),(v,y)\} \cap P^- \neq \emptyset\}.$$

The following holds.

Proposition 3.5 [Bonchi et al., 2011]   *The chromatic number $\chi(\widehat{G})$ of $\widehat{G}$ is equal to the minimum number of cluster labels required by a zero-cost solution to the (b, I, k) problem with input (V, s).*

Proof. We observe that a color in $\widehat{G}$ corresponds to a cluster in our problem. The colors are assigned to pairs of objects in V, which ensures that the positive pairs P^+ are satisfied. On the other hand, the constraint of having a proper coloring, ensures that the negative pairs P^- are also satisfied. Thus, a proper coloring on $\widehat{G}$ corresponds to a zero-cost solution on our problem. □

Although the previous result is theoretically interesting, it has limited practical relevance, as we are interested in minimizing the error given a specific number of clusters. To make the connection practically useful, we would need to relax the GRAPH-COLORING problem, so that it allows for a less strict definition of coloring. Namely, we would like that, for a certain cost, colorings may allow the following relaxations: (i) $(u,v) \in E$ not necessarily implies $c(u) \neq c(v)$—corresponding to violations on P^-; and (ii) vertices may be left uncolored—corresponding to violations on P^+.

3.1.4   CONNECTION WITH INTERSECTION REPRESENTATION

As observed by Tsourakakis [2015] and Li et al. [2017], when the set-intersection indicator is used, i.e., for problems of the form (b, I, k), OVERLAPPING-CORRELATION-CLUSTERING reduces to an instance of INTERSECTION-REPRESENTATION, a well-established problem in graph theory [Erdös et al., 1966]. An intersection representation of a finite, undirected graph $G = (V(G), E(G))$ is an assignment of subsets $\mathcal{I}_u$ of a finite, sufficiently large ground set $\mathcal{F}$, to vertices $u \in V$ such that $(u,v) \in E$ if and only if $\mathcal{I}_u \cap \mathcal{I}_v \neq \emptyset$. The smallest cardinality of the ground set $\mathcal{F}$ needed to properly represent the graph is known as the *intersection number of the graph*. It is known that the intersection number of a graph equals its *edge clique-cover number*, i.e., the smallest number of cliques in the graph needed to cover all the edges in the graph [Erdös et al., 1966]. It is easy to see that, given an intersection representation of the graph, the set of vertices that are assigned to a particular clique may be seen as sharing one feature. This is why the intersection representation of a graph is often used for latent feature inference.

An example of an intersection representation of a graph over the smallest ground set $\mathcal{F} = \{1, 2, 3\}$ is shown in Figure 3.1.

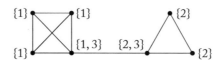

Figure 3.1: An intersection representation of a graph using three features $\{1, 2, 3\}$.

3.1.5   CONNECTION WITH DIMENSIONALITY REDUCTION

Given a set of points in a high-dimensional space, the goal of dimensionality reduction is to map each point x in the original space to a point $h(x)$ in a space of lower dimensionality, so that for any pair x, y of points, their $d(x, y)$ distance in the high-dimensional space is preserved as well as possible by the $d(h(x), h(y))$ distance in the lower-dimensional space. Dimensionality reduction is a problem that has been studied, among other areas, in theory, data mining, and machine learning, and has many applications, for example, in proximity search, feature selection, component analysis, visualization, and more. The connection between dimensionality reduction and the OVERLAPPING-CORRELATION-CLUSTERING problem is apparent. However, the difference is that dimensionality-reduction methods are typically defined for geometric spaces. Alternatively, they operate by hashing high-dimensional or complex objects in a way that similar objects have high collision probability [Broder et al., 1998, Indyk and Motwani, 1998].

A specific approach to dimensionality reduction, namely nonnegative matrix factorization, is discussed more in detail next.

3.1.6   CONNECTION WITH NONNEGATIVE MATRIX FACTORIZATION

In the nonnegative matrix factorization (NMF) problem [Lee and Seung, 2001], the goal is to decompose a given matrix $\mathbf{A} \in \mathbb{R}^{m \times n}$ into the product of two nonnegative matrices $\mathbf{W} \in \mathbb{R}^{m \times k}$ and $\mathbf{H} \in \mathbb{R}^{k \times n}$, so that the product $\mathbf{WH}$ approximates as well as possible the original matrix A. The quality of the decomposition is typically measured with the Frobenious matrix norm $||\mathbf{A} - \mathbf{WH}||_F^2$. It is also assumed that the dimension k is much smaller than the dimensions m and n. In the symmetric version of the nonnegative matrix factorization problem (SNMF) [Ding et al., 2005, He et al., 2011], the input matrix A is symmetric, i.e., $\mathbf{A} \in \mathbb{R}^{n \times n}$ and $\mathbf{A} = \mathbf{A}^T$, and the goal is to find a nonnegative matrix $\mathbf{X} \in \mathbb{R}^{k \times n}$ so that $\mathbf{A} \approx \mathbf{X}^T \mathbf{X}$.

The SNMF problem has many similarities with the OVERLAPPING-CORRELATION-CLUSTERING problem. Consider that the input matrix $\mathbf{A} = [a_{ij}]$ represents the pairwise object similarities, i.e., $a_{ij} = s(i, j)$. The factor matrix $\mathbf{X}$ can be viewed as a mapping from each object $i \in V$ to a k-dimensional vector, namely, to the i-th column $\mathbf{x}_i$ of the matrix $\mathbf{X} = [\mathbf{x}_1 \ldots \mathbf{x}_n]$. Should the matrix $\mathbf{X}$ have 0-1 values, the mapping of $i \in V$ to $\mathbf{x}_i$ can be interpreted as a membership vector to k clusters, and thus, the solution to the SNMF problem has a natural interpretation as overlapping clustering. Furthermore, the objective function of the SNMF problem is $||\mathbf{A} - \mathbf{X}^T \mathbf{X}||_F^2 = \sum_{i,j} |a_{ij} - \langle \mathbf{x}_i, \mathbf{x}_j \rangle|^2$, where $\langle \cdot, \cdot \rangle$ denotes the dot-product operation. A similar problem definition has been considered in the statistics literature, with the goal of modeling graphs using dot-product representations [Scheinerman and Tucker, 2010].

On the other hand, there are also many differences of the approaches described above with the OVERLAPPING-CORRELATION-CLUSTERING problem. First, in OVERLAPPING-COR-RELATION-CLUSTERING instead of the dot-product between representation vectors, the measures of set-intersection and Jaccard are used. Second, in the above-mentioned problems, there is no integrality constraint for the entries of the solution matrix $\mathbf{X}$, and thus, there is no immediate interpretation of the solution as overlapping clustering. One way to obtain integral solutions is to *round* the fractional solutions using a threshold parameter. This approach is actually explored in Bonchi et al. [2011, 2013b], where it is shown that, compared against a ground-truth, OVER-LAPPING-CORRELATION-CLUSTERING algorithms outperform the SNMF-based algorithms for a wide range of the rounding threshold parameter.

3.1.7 CONNECTION WITH LATENT FEATURES LEARNING

Latent feature models for graphs aim at explaining connections (i.e., edges) by learning a set of features for each vertex. Inference of latent network features is useful in several analysis tasks, such as community detection and link prediction, and several application domains, such as social networks, protein complexes, and gene regulatory modules.

Two works have built on top of Bonchi et al. [2011]'s overlapping-correlation-clustering framework to derive models for latent features learning [Dau and Milenkovic, 2017, Tsourakakis, 2015]. Tsourakakis [2015] proposes a probabilistic latent feature model for graphs under the assumption that each vertex has k latent binary features, and two vertices form a connection with higher probability if they share more features. Maximizing the log-likelihood function for this simple model is shown to be **NP**-hard, as it subsumes as a special instance the OVERLAPPING-CORRELATION-CLUSTERING problem. Then, Tsourakakis [2015] provides a rapidly mixing Markov chain for learning the latent features. While doing this, Tsourakakis [2015] also shows that the algorithms of Bonchi et al. [2011] (presented in the next section) can be seen as a special case of the mixing Markov chain, essentially corresponding to a deterministic hill-climb algorithm. This also explains the main drawback of Bonchi et al. [2011]'s methods, namely being prone to local optima.

Dau and Milenkovic [2017] extend the combinatorial variant of the model studied by Bonchi et al. [2011] and Tsourakakis [2015] to a much more general setting, by using Boolean functions of features that can express more complicated interactions among vertices. For instance, suppose that there are three different types of features, namely "Family member," "City," and "Hobby." The Boolean function $f(x1, x2, x3) = x1 \lor (x2 \land x3)$ can be used to express the connection rule that two people are Facebook friends if and only if either they are family members or they have lived in at least one common city and shared at least one common hobby. As such, it asserts that the "Family member" feature is more relevant than either of the "City" or "Hobby" features.

Algorithm 3.8 LocalSearch [Bonchi et al., 2011]

1: initialize ℓ to a valid labeling;
2: **while** $C_{\mathrm{occ}}(V, \ell)$ decreases **do**
3:      **for** each $v \in V$ **do**
4:          find the label set L that minimizes $C_{v,p}(L \mid \ell)$;
5:          update ℓ so that $\ell(v) = L$;
6:      **end for**
7: **end while**
8: **return** ℓ

3.1.8   ALGORITHMS

A typical approach for multivariate optimization problems is to iteratively find the optimal value for one variable *given* values for the remaining variables. The global solution is found by repeatedly optimizing each of the variables in turn until the objective function value no longer improves. In most cases such a method will converge to a local optimum. Bonchi et al. [2011] propose a local-search algorithm for the OVERLAPPING-CORRELATION-CLUSTERING problem that falls into this framework. At the core of the proposal is an efficient method for finding a good labeling of a single object given a fixed labeling of the other objects, with the guarantee that the value of Equation (3.1) is non-increasing with respect to such optimization steps.

First, Bonchi et al. [2011] observe that the cost of Equation (3.1) can be rewritten as

$$C_{\mathrm{occ}}(V, \ell) = \frac{1}{2} \sum_{v \in V} \sum_{u \in V \setminus \{v\}} |H(\ell(v), \ell(u)) - s(v, u)|$$

$$= \frac{1}{2} \sum_{v \in V} C_{v,p}(\ell(v) \mid \ell),$$

where

$$C_{v,p}(\ell(v) \mid \ell) = \sum_{u \in V \setminus \{v\}} |H(\ell(v), \ell(u)) - s(v, u)| \tag{3.2}$$

expresses the error incurred by vertex v when it has the labels $\ell(v)$, and the remaining vertices are labeled according to ℓ. The subscript p in $C_{v,p}$ serves to remind us that the set $\ell(v)$ should have at most p labels.

The proposed local-search strategy is summarized in Algorithm 3.8. Line 4 is the step where LocalSearch attempts to find an optimal set of labels for an object v by solving Equation (3.2). This is also the step in which the framework proposed by Bonchi et al. [2011] differentiates between the measures of Jaccard coefficient and set-intersection.

For the case of the Jaccard coefficient, Bonchi et al. [2011] provide the following precise problem formulation.

Problem 3.6 (Jaccard-Triangulation) Consider the set $\{\langle S_j, z_j \rangle\}_{j=1\ldots n}$, where S_j are subsets of a ground set $U = \{1, \ldots, k\}$, and z_j are fractional numbers in the interval $[0, 1]$. The task is to find a set $X \subseteq U$ that minimizes the distance

$$d(X, \{\langle S_j, z_j \rangle\}_{j=1\ldots n}) = \sum_{j=1}^{n} |J(X, S_j) - z_j|. \qquad (3.3)$$

The intuition behind Equation (3.3) is that we are given sets S_j and "target similarities" z_j and we want to find a set whose Jaccard coefficient with each set S_j is as close as possible to the target similarity z_j. A moment's thought can convince us that Equation (3.3) corresponds exactly to the error term $C_{v,p}(\ell(v) \mid \ell)$ defined in Equation (3.2), and thus, in the local-improvement step of the LocalSearch algorithm.

An interesting problem related to Jaccard-Triangulation is the problem of finding the *Jaccard median* [Chierichetti et al., 2010]. More precisely, the JACCARD-MEDIAN problem is a special case of the Jaccard-Triangulation problem, where all similarities z_j are equal to 1. Since Jaccard-Triangulation is a generalization of the Jaccard-median problem that has been proven to be **NP**-hard [Chierichetti et al., 2010], Jaccard-Triangulation is **NP**-hard as well.

We next discuss Bonchi et al. [2011]'s algorithm for the Jaccard-Triangulation problem. The idea is to introduce a variable x_i for every element $i \in U$. The variable x_i indicates if element i belongs in the solution set X. In particular, $x_i = 1$ if $i \in X$ and $x_i = 0$ otherwise. We then assume that the size of set X is t, that is,

$$\sum_{i \in U} x_i - t = 0. \qquad (3.4)$$

Now, given a set S_j with target similarity z_j we want to obtain $J(X, S_j) = z_j$, for all $j = 1, \ldots n$, or

$$J(X, S_j) = \frac{\sum_{i \in S_j} x_i}{|S_j| + t - \sum_{i \in S_j} x_i} = z_j,$$

which is equivalent to

$$(1 + z_j) \sum_{i \in S_j} x_i - z_j t = z_j |S_j|, \qquad (3.5)$$

and we have one equation of type (3.5) for each pair $\langle S_j, z_j \rangle$. We observe that Equations (3.4) and (3.5) are linear with respect to the unknowns x_i and t. On the other hand, the variables x_i and t take integral values, which implies that the system of Equations (3.4) and (3.5) cannot be solved efficiently. Instead, Bonchi et al. [2011] propose to relax the integrality constraints to nonnegativity constraints $x_i, t \geq 0$ and solve the above system by applying a nonnegative least-squares optimization method (NNLS), obtaining estimates for the variables x_i and t.

The solution obtained from the NNLS solver has two drawbacks: (i) it does not incorporate the constraint of having at most p labels, and, more importantly, (ii) it does not have a clear interpretation as a set X, since the variables x_i may take any nonnegative value, not only 0-1. Bonchi et al. [2011] address both of these problems with a greedy post-processing of the fractional solution: the variables x_i are sorted in decreasing order, breaking ties arbitrarily. Then a set X_q is produced by setting the first q variables x_i to 1 and the rest to 0 (with q varying from 1 to p). Out of the p different sets X_q obtained, the one that minimizes the cost $d(X_q, \{\langle S_j, z_j \rangle\})$ is returned as the solution to the Jaccard-Triangulation problem.

For the case of the set-intersection function I, Bonchi et al. [2011] formulate the problem underlying the local-improvement step of the LocalSearch algorithm (Line 4 of Algorithm 3.8) as follows.

Problem 3.7   (Hit-n-Miss)   Let C be a collection of n tuples of the from $\langle S_j, h_j, m_j \rangle$, with $j = 1 \ldots n$, where S_j are subsets of a ground set $U = \{1, \ldots, k\}$, while h_j and m_j are nonnegative numbers. A set $X \subseteq U$ partition C in $C_X = \{S_j \mid I(X, S_j) = 1\}$ and $C_{\bar{X}} = \{S_j \mid I(X, S_j) = 0\}$. The task is a find a set X in order to minimize the distance

$$d(X, \{\langle S_j, h_j, m_j \rangle\}) = \sum_{j \mid S_j \in C_X} h_j + \sum_{j \mid S_j \in C_{\bar{X}}} m_j. \qquad (3.6)$$

Once again, we should be able to verify that Equation (3.6) corresponds to the cost $C_{v,p}(\ell(v) \mid \ell)$ defined by Equation (3.2) in the case that the cluster-label similarity function H is the set-intersection function I.

The Hit-n-Miss problem is related to set-cover-type problems. As in SET-COVER, we are given a collection C of sets S_j. Each set is accompanied by two penalty scores, a *hit* penalty p_j and a *miss* penalty n_j. The goal is to find a new set X in order to either hit or miss the sets S_j, as dictated by their penalty scores h_j and m_j. In particular, for each set S_j that X hits, we pay its hit penalty h_j, while for each set S_j that X misses we pay its miss penalty m_j. The Hit-n-Miss problem is isomorphic to the POSITIVE-NEGATIVE-PARTIAL-SET-COVER problem, studied by Miettinen [2008], who showed that the problem is not approximable within a constant factor, but it admits an $\mathcal{O}(\sqrt{n} \log n)$ approximation.

Bonchi et al. [2011] tackle the Hit-n-Miss problem with a simple greedy strategy: starting from $X_0 = \emptyset$, let X_t the current solution and let $A = U \setminus X_t$ be the set of currently available items (cluster labels). Then for the next step of the greedy pick the item i from the set of available items A that yields the lowest distance cost, evaluated as $d(X_t \cup \{i\}, \{\langle S_j, h_j, m_j \rangle\})$. The process terminates when there is no further decrease in the cost or when the maximum number of cluster labels allowed is reached, i.e., $t = p$.

A distributed version of the algorithms introduced in this section, implemented on a *map-reduce* architecture [Dean and Ghemawat, 2008], that can be used to cluster really large problem instances is provided in Bonchi et al. [2013b].

Andrade et al. [2014] proposed a genetic approach, named Biased Random Key Genetic Algorithms (BRKGA), to solve OVERLAPPING-CORRELATION-CLUSTERING problem. In their experiments, Andrade et al. [2014] show that BRKGA is effective at finding good solutions, often outperforming Bonchi et al. [2013b], when using the Jaccard index in the objective function. While for set intersection, Bonchi et al. [2013b] achieve better results than other algorithms in most scenarios. One of the main drawbacks of BRKGA is its high running time.

3.2   CORRELATION CLUSTERING WITH LOCAL OBJECTIVES

The standard objective function of correlation clustering is a global one, as it aims at minimizing (maximizing) the overall disagreements (agreements) of a clustering. Puleo and Milenkovic [2018] depart from this view and initiate the study of *local objectives* for correlation clustering, where the disagreements (agreements) are considered locally, at the level of a single vertex, and the goal is to minimize (maximize) an aggregation function of the disagreements (agreements) for edges incident on a single vertex. Such a generalization of correlation clustering may be useful in several application scenarios. As an example, using the ℓ_∞-norm as an aggregation function in a MIN-DISAGREE formulation leads to a *minimax* variant of correlation clustering, whose goal is to minimize the error encountered at the worst-off vertex in the clustering. This finds easy application in detecting communities, such as gene, social network, or voter communities, in which no *antagonists* (entities exhibiting properties inconsistent with a large number of members of the community) are allowed. Another application is quality control on individual vertices within the various clusters, which may be relevant in biclustering contexts such as collaborative filtering for recommender systems, where minimum-quality recommendations have to be ensured for each user in a given category. Other examples include multifeature classification and reconciliation, as well as identification of cancer driver gene communities in cancer bioinformatics.

Formally, the local-objective-aware generalization of correlation clustering defined in Puleo and Milenkovic [2018] is as follows. Given a weighted signed graph $G = (V, E, w, \mathcal{L})$ with n vertices, and a clustering $\ell : V \to \mathbb{N}$ of V, let v-err(ℓ) and v-agree(ℓ) be two n-dimensional real-valued vectors containing, for every vertex $u \in V$, the objective-function error and gain achieved in correspondence of u, respectively. Specifically, every entry v-err(ℓ)$_u$ and v-agree$(\ell)_u$, for all $u \in V$, is defined, respectively, as:

$$\text{v-err}(\ell)_u = \sum_{\substack{v \in V:(u,v) \in E, \\ \ell(u)=\ell(v)}} \frac{1 - \mathcal{L}(u,v)}{2} w(u,v) + \sum_{\substack{v \in V:(u,v) \in E, \\ \ell(u) \neq \ell(v)}} \frac{1 + \mathcal{L}(u,v)}{2} w(u,v), \qquad (3.7)$$

$$\text{v-agree}(\ell)_u = \sum_{\substack{v \in V:(u,v) \in E, \\ \ell(u)=\ell(v)}} \frac{1 + \mathcal{L}(u,v)}{2} w(u,v) + \sum_{\substack{v \in V:(u,v) \in E, \\ \ell(u) \neq \ell(v)}} \frac{1 - \mathcal{L}(u,v)}{2} w(u,v). \qquad (3.8)$$

Puleo and Milenkovic [2018] formulate the MIN-DISAGREE-WITH-VERTEX-LOCAL-GUARANTEES and MAX-AGREE-WITH-VERTEX-LOCAL-GUARANTEES problems as follows.

Problem 3.8 (Min-Disagree-with-Vertex-Local-Guarantees) Given a weighted signed graph $G = (V, E, w, \mathcal{L})$, where V is a set of n vertices, $E \subseteq \binom{V}{2}$ is a set of m edges, $w : E \to \mathbb{R}^+$ is a function assigning a positive weight to each edge, and $\mathcal{L} : E \to \{-1, +1\}$ is a function labeling an edge as either positive or negative, and a function $f : \mathbb{R}^n_{\geq 0} \to \mathbb{R}$, find a clustering $\ell : V \to \mathbb{N}$ so as to minimize

$$f(\text{v-err}(\ell)). \qquad (3.9)$$

Problem 3.9 (Max-Agree-with-Vertex-Local-Guarantees) Given a weighted signed graph $G = (V, E, w, \mathcal{L})$, where V is a set of n vertices, $E \subseteq \binom{V}{2}$ is a set of m edges, $w : E \to \mathbb{R}^+$ is a function assigning a positive weight to each edge, and $\mathcal{L} : E \to \{-1, +1\}$ is a function labeling an edge as either positive or negative, and a function $f : \mathbb{R}^n_{\geq 0} \to \mathbb{R}$, find a clustering $\ell : V \to \mathbb{N}$ so as to maximize

$$f(\text{v-agree}(\ell)). \qquad (3.10)$$

It is easy to see that setting f to the ℓ_1-norm, MIN-DISAGREE-WITH-VERTEX-LOCAL-GUARANTEES and MAX-AGREE-WITH-VERTEX-LOCAL-GUARANTEES reduce to the traditional MIN-DISAGREE and MAX-AGREE problems, respectively.

Similarly in spirit to Puleo and Milenkovic [2018]'s problems, Ahmadi et al. [2019] devise a *cluster-wise* variant of correlation clustering with local objectives, where the goal is to minimize the maximum error observed in a cluster. This captures the desideratum of finding balanced clusters, which may arise in problems such as image segmentation and community detection. Specifically, given a weighted signed graph $G = (V, E, w, \mathcal{L})$ with n vertices, and a clustering $\ell : V \to \mathbb{N}$ of V, let c-err$(\ell)_c$ denote the cost paid by the correlation-clustering objective function in correspondence of a cluster c of ℓ, i.e., the sum of the weights of negative edges within c plus the sum of the weights of positive edges with an endpoint in c and the other endpoint in a cluster other than c:

$$\text{c-err}(\ell)_c = \sum_{\substack{(u,v)\in E, \\ \ell(u)=c, \\ \ell(v)=c}} \frac{1 - \mathcal{L}(u,v)}{2} w(u,v) \; + \sum_{\substack{(u,v)\in E, \\ \ell(u)=c, \\ \ell(v)\neq c}} \frac{1 + \mathcal{L}(u,v)}{2} w(u,v). \tag{3.11}$$

The problem introduced by Ahmadi et al. [2019] is the following.

Problem 3.10   (Min-Disagree-with-Cluster-Local-Guarantees)   Given a weighted signed graph $G = (V, E, w, \mathcal{L})$, where V is a set of n vertices, $E \subseteq \binom{V}{2}$ is a set of m edges, $w : E \to \mathbb{R}^+$ is a function assigning a positive weight to each edge, and $\mathcal{L} : E \to \{-1, +1\}$ is a function labeling an edge as either positive or negative, find a clustering $\ell : V \to \mathbb{N}$ so as to minimize

$$\max_{c \in \{\ell(u)|u\in V\}} \text{c-err}(cl)_c. \tag{3.12}$$

In the remainder of this section we discuss the main theoretical results and algorithms devised for the above three problems.

3.2.1   VERTEX-WISE FORMULATIONS

Hardness Results

Puleo and Milenkovic [2018] show that the minimax variant of their local-objective-aware correlation-clustering problem—that is, MIN-DISAGREE-WITH-VERTEX-LOCAL-GUARANTEES with f being set to the ℓ_∞-norm—is **NP**-hard, even on unweighted complete graphs and unweighted complete bipartite graphs. This clearly implies **NP**-hardness of both MIN-DISAGREE-WITH-VERTEX-LOCAL-GUARANTEES and MAX-AGREE-WITH-VERTEX-LOCAL-GUARANTEES in general. Puleo and Milenkovic [2018]'s **NP**-hardness proof for complete graphs employs a reduction from the PARTITION-INTO-TRIANGLES problem on 4-regular graphs due to van Rooij et al. [2013]. The PARTITION-INTO-TRIANGLES problem is a decision problem asking whether the vertex set of a graph $G = (V, E)$, with $|V| = 3t$ for some integer t, can be partitioned into t sets $V_1, \ldots, V_t$ such that V_i induces a triangle in G, for all $i \in [1 \ldots t]$. Puleo and Milenkovic [2018]'s reduction mimics the proof given by Bansal et al. [2004] for the basic formulation of correlation clustering and reported in Section 1.3.

As far as the bipartite-graph case, Puleo and Milenkovic [2018] give instead a reduction from the 3-Cover (decision) problem: given a universe U of elements, and a family $\mathcal{S}$ of subsets of U, with every $S \in \mathcal{S}$ having size $|S| = 3$, is there a subfamily $\mathcal{S}' \subseteq \mathcal{S}$ such that each $x \in U$ lies in exactly one element of $\mathcal{S}'$? The 3-Cover problem is well-known to be **NP**-complete [Garey and Johnson, 1979]. The reduction devised by Puleo and Milenkovic [2018] resembles the one by Amit [2004] for the **NP**-hardness of the standard formulation of biclustering. However, it also requires significant modifications to accommodate the specific objective function of the MIN-DISAGREE-WITH-VERTEX-LOCAL-GUARANTEES problem. All the details can be found in Puleo and Milenkovic [2018, Appendix D].

Disagreement Minimization on Complete (Unweighted) Graphs
A 48-approximation algorithm. Puleo and Milenkovic [2018] devise a 48-approximation algorithm for MIN-DISAGREE-WITH-VERTEX-LOCAL-GUARANTEES on unweighted complete graphs and for aggregation functions f satisfying the following two conditions.

Assumption 1 $f : \mathbb{R}^n_{\geq 0} \to \mathbb{R}$ *has the following properties:*

1. *(Monotonicity) For any $\vec{x}, \vec{y} \in \mathbb{R}^n_{\geq 0}$, if $\vec{x} \leq \vec{y}$ then $f(\vec{x}) \leq f(\vec{y})$.*

2. *(Scaling) $f(\alpha \vec{x}) \leq \alpha f(\vec{x})$, for any $\alpha \geq 0$ and $\vec{x} \in \mathbb{R}^n_{\geq 0}$.*

Given an unweighted signed graph $G = (V, E, \mathcal{L})$, let a *fractional clustering* $\vec{c}$ of G be an n-dimensional vector indexed by $\binom{V}{2}$ such that $\vec{c}_{uv} \in [0, 1]$, for all $uv \in \binom{V}{2}$, and $\vec{c}_{uw} \leq \vec{c}_{uv} + \vec{c}_{vw}$, for all distinct $u, v, w \in V$. A fractional clustering c may result, e.g., from the solution of a convex-programming relaxation of the MIN-DISAGREE-WITH-VERTEX-LOCAL-GUARANTEES problem instance at hand. An entry $\vec{c}_{uv}$ of a fractional clustering can be viewed as a distance between u and v (more precisely, a bounded semimetric of u and v without the identity of indiscernibles property), with the convention that $\vec{c}_{uu} = 0$, for all u. The main contribution of Puleo and Milenkovic [2018] is a rounding algorithm to transform an arbitrary fractional clustering $\vec{c}$ of a complete unweighted signed graph G into a hard clustering ℓ of G such that

$$\text{v-err}(\ell)_u \leq W \text{v-err}(\vec{c})_u, \text{ for all } u \in V \text{ and some constant } W > 1.$$

Under Assumption 1, this property guarantees that $f(\text{v-err}(\ell)) \leq W f(\text{v-err}(\vec{c}))$. Therefore, if one can get a fractional clustering $\vec{c}$ that is a $(1 + \varepsilon)$-approximation for the corresponding fractional version of the MIN-DISAGREE-WITH-VERTEX-LOCAL-GUARANTEES problem instance, applying Puleo and Milenkovic [2018]'s rounding algorithm to such a $\vec{c}$ guarantees a $W(1 + \varepsilon)$-approximation for the original MIN-DISAGREE-WITH-VERTEX-LOCAL-GUARANTEES problem

---

Algorithm 3.9 Rounding-Min-Disagree-with-Vertex-Local-Guarantees [Puleo and Milenkovic, 2018]

Input: A complete unweighted signed graph $G = (V, E, \mathcal{L})$, a fractional clustering $\vec{c}$ of G, parameters α, γ, with $0 < \gamma < \alpha < \frac{1}{2}$

Output: A hard clustering ℓ of G

1: $S \leftarrow V$
2: **while** $S \neq \emptyset$ **do**
3:      For each $v \in S$, let $T_v = \{w \in S : \vec{c}_{vw} \leq \alpha\}$ and $T_v^* = \{w \in S : \vec{c}_{uw} \leq \gamma\}$
4:      Choose a pivot vertex $u \in S$ that maximizes $|T_u^*|$
5:      **if** $\sum_{w \in T_u} \vec{c}_{uw} \geq \alpha|T_u|/2$ **then**
6:          $T \leftarrow T_u$
7:      **else**
8:          $T \leftarrow \{u\}$
9:      **end if**
10:     Add cluster T to the output ℓ clustering
11:     $S \leftarrow S \setminus T$
12: **end while**

---

instance. If f is linear, the convex-programming relaxation of the Min-Disagree-with-Vertex-Local-Guarantees problem instance at hand can be solved exactly in polynomial time, thus meaning that $\varepsilon = 0$ and the approximation factor W of the rounding algorithm carries over as is to the actual Min-Disagree-with-Vertex-Local-Guarantees problem. Another tractable case is when f is convex on $\mathbb{R}^n_{\geq 0}$. In this case, in fact, the composite function $f \circ \text{v-err}$ is convex as well, which, along with the fact that the constraints defining a fractional clustering are linear inequalities in the variables $\vec{c}_{uv}$ of the problem, make fractional Min-Disagree-with-Vertex-Local-Guarantees solvable in polynomial time by means of standard techniques from convex optimization [Boyd and Vandenberghe, 2004]. A notable class of convex objective functions obeying Assumption 1 is the class of ℓ_p-norms, i.e., the class of functions $f : \mathbb{R}^n_{\geq 0} \to \mathbb{R}$ of the form $f(\vec{x}) = \left(\sum_{i=1}^n |\vec{x}_i|^p\right)^{1/p}$, for all $p \geq 1$. However, it is worth emphasizing that the correctness of Puleo and Milenkovic [2018]'s result does not depend on the convexity of f, but only on the properties listed in Assumption 1. Indeed, if f is nonconvex but obeys Assumption 1, and a "good" fractional clustering $\vec{c}$ (with constant-factor approximation guarantee) can be produced by some means, still Puleo and Milenkovic [2018]'s rounding algorithm yields a hard clustering ℓ achieving a constant-factor approximation guarantee for the original Min-Disagree-with-Vertex-Local-Guarantees problem instance.

The rounding algorithm devised by Puleo and Milenkovic [2018] is shown as Algorithm 3.9. It resembles the rounding algorithm at the basis of the 4-approximation algorithm for Min-Disagree devised in Charikar et al. [2005]. The main difference between Algorithm 3.9 and the algorithm of Charikar et al. [2005] lies in the strategy of choosing a pivot vertex:

in Charikar et al. [2005] a pivot vertex is chosen arbitrarily, while in Algorithm 3.9 it is chosen by maximizing the number of vertices that, according to the fractional clustering, are at distance no more than a parameter γ from the selected pivot. Furthermore, the algorithm of Charikar et al. [2005] uses a cutoff of $\frac{1}{2}$ to form candidate clusters, while Algorithm 3.9 has a parameter $\alpha \in (0, \frac{1}{2})$ to set that cutoff. Despite their (apparent) simplicity, these two modifications are enough to make the theoretical analysis of Algorithm 3.9 complex and with a number of additional non-trivial technical challenges with respect to the analysis of Charikar et al. [2005]'s rounding algorithm. The main reason is that, similarly to Charikar et al. [2005], Algorithm 3.9 has to pay for the cluster-cost of the errors incurred by "charging" the cost of these errors to the LP-costs of the fractional clustering. The difference lies in the fact that Algorithm 3.9 incurs *local* errors: for each vertex u, *all* clustering errors incident to u must be taken into account by charging to the LP cost incident to u. In particular, every clustering error must now be paid for at *each* of its endpoints, while in Charikar et al. [2005] it is enough to pay for each clustering error at *one* of its endpoints. For edges which cross between a cluster and its complement, this requires a different analysis at each endpoint, a difficulty that is not present in Charikar et al. [2005]. Ultimately, Puleo and Milenkovic [2018] provide the following result.

Theorem 3.11　**[Puleo and Milenkovic, 2018].**　*Under Assumption 1, and given a fractional clustering $\vec{c}$ that is a solution of the corresponding convex-programming relaxation of the* Min-Disagree-with-Vertex-Local-Guarantees *problem instance, Algorithm 3.9 achieves a* $\max\{c_1, c_2, c_3\}$*-approximation factor for* Min-Disagree-with-Vertex-Local-Guarantees *on complete graphs, where*

$$c_1 = \frac{1}{(1 - 2k_3)(k_3 - k_2)\alpha} + \frac{1}{1 - 2\alpha} + \frac{1}{k_1\alpha - \gamma},$$

$$c_2 = \frac{1}{(1 - 2k_3)(k_3 - k_2)\alpha} + \max\left\{\frac{1}{(1 - k_1)\alpha}, \frac{1}{\gamma}\right\},$$

$$c_3 = \frac{1}{(1 - 2k_3)(k_3 - k_2)\alpha} + \frac{1}{k_2\alpha},$$

and k_1, k_2, k_3 are three constants, with $\frac{1}{2} < k_1 < 1$, $0 < 2k_2 \leq k_3 < \frac{1}{2}$, and $k_1\alpha > \gamma$, $k_2\alpha \leq 1 - 2\alpha$.

Puleo and Milenkovic [2018] point out that analytically choosing the best parameter values to make the approximation ratio in Theorem 3.11 as small as possible is a hard task. Instead, they provide the following numerical solution

$$\alpha = 0.465744, \qquad \gamma = 0.0887449,$$
$$k_1 = 0.767566, \qquad k_2 = 0.117219, \qquad k_3 = 0.308433,$$

which yields an overall approximation factor of roughly 48.

Algorithm 3.10 Rounding-Min-Disagree-with-Vertex-Local-Guarantees [Charikar et al., 2017]

Input: A complete unweighted signed graph $G = (V, E, \mathcal{L})$, a fractional clustering $\vec{c}$ of G
Output: A hard clustering ℓ of G

  1: $S \leftarrow V$
  2: **while** $S \neq \emptyset$ **do**
  3:     For each $v \in S$, let $T_v = \{w \in S : \vec{c}_{vw} < \frac{3}{7}\}$ and $T_v^* = \{w \in S : \vec{c}_{uw} < \frac{1}{7}\}$
  4:     Choose a pivot vertex $u \in S$ that maximizes $|T_u^*|$
  5:     Add cluster T_u to the output ℓ clustering
  6:     $S \leftarrow S \setminus T_u$
  7: **end while**

Algorithm 3.11 Rounding-Min-Disagree-with-Vertex-Local-Guarantees [Kalhan et al., 2019]

Input: A complete unweighted signed graph $G = (V, E, \mathcal{L})$, a fractional clustering $\vec{c}$ of G
Output: A hard clustering ℓ of G

  1: $S \leftarrow V$
  2: **while** $S \neq \emptyset$ **do**
  3:     For each $v \in S$, let $T_v = \{w \in S : \vec{c}_{vw} \leq \frac{2}{5}\}$ and $T_v^* = \{w \in S : \vec{c}_{vw} \leq \frac{1}{5}\}$
  4:     Choose a pivot vertex $u \in S$ that maximizes $\sum_{w \in T_u^*} \left(\frac{1}{5} - \vec{c}_{uw}\right)$
  5:     Add cluster T_u to the output ℓ clustering
  6:     $S \leftarrow S \setminus T_u$
  7: **end while**

A 7-approximation algorithm.   Charikar et al. [2017] devise an alternative rounding algorithm to the one of Puleo and Milenkovic [2018] (Algorithm 3.9). Charikar et al. [2017]'s algorithm, whose pseudocode is shown as Algorithm 3.10, resembles Puleo and Milenkovic [2018]'s one. The main difference is that Charikar et al. [2017]'s algorithm sets parameters α and γ to $\frac{3}{7}$ and $\frac{1}{7}$, respectively, and it *always* builds a cluster corresponding to a sphere of radius $\frac{3}{7}$ around the chosen u pivot. Instead, Puleo and Milenkovic [2018]'s algorithm outputs either a singleton cluster (containing the u pivot only) or some other sphere around u (the average distance within the sphere determines which one of the two options is chosen). Restricting the algorithm's choice allows for obtaining a simpler algorithm, as well as an improved approximation guarantee, from 48 to 7.

Theorem 3.12   Charikar et al. [2017].   *Under Assumption 1, and given a fractional clustering $\vec{c}$ that is a solution of the corresponding convex-programming relaxation of the* MIN-DISAGREE-WITH-VERTEX-LOCAL-GUARANTEES *problem instance, Algorithm 3.10 achieves a 7-approximation factor for* MIN-DISAGREE-WITH-VERTEX-LOCAL-GUARANTEES *on complete graphs.*

Algorithm 3.12 Bipartite-Rounding-Min-Disagree-with-Vertex-Local-Guarantees
[Puleo and Milenkovic, 2018]

Input: A complete unweighted signed bipartite graph $G = (V_1, V_2, \mathcal{L})$, a fractional clustering $\vec{c}$ of G, parameters α, γ, with $0 < \gamma < \alpha < \frac{1}{2}$

Output: A hard clustering ℓ of G

 1: $S \leftarrow V$
 2: **while** $V_1 \cap S \neq \emptyset$ **do**
 3:     For each $v \in V_1 \cap S$, let $T_v = \{w \in S : \vec{c}_{vw} \leq \alpha\}$ and let $T_v^* = \{w \in V_2 \cap S : \vec{c}_{vw} \leq \gamma\}$
 4:     Choose a pivot vertex $u \in V_1 \cap S$ that maximizes $|T_u^*|$
 5:     **if** $\sum_{w \in V_2 \cap T_u} \vec{c}_{uw} \geq \alpha |V_2 \cap T_u|/2$ **then**
 6:         $T \leftarrow \{u\}$
 7:     **else**
 8:         $T \leftarrow T_u$
 9:     **end if**
10:     Add cluster T to the output ℓ clustering
11: **end while**
12: Add every vertex in S (if any) as a singleton cluster to the output ℓ clustering

---

A 5-approximation algorithm.   Kalhan et al. [2019] provide yet another rounding algorithm, which is outlined as Algorithm 3.11 and is shown to achieve a 5-approximation guarantee when an ℓ_p-norm, for any $p \geq 1$, is employed as an aggregation function f.

Theorem 3.13   [Kalhan et al., 2019]. *For an aggregation function f corresponding to an ℓ_p-norm, for some $p \geq 1$, and given a fractional clustering $\vec{c}$ that is a solution of the corresponding convex-programming relaxation of the* MIN-DISAGREE-WITH-VERTEX-LOCAL-GUARANTEES *problem instance, Algorithm 3.11 achieves a 5-approximation factor for* MIN-DISAGREE-WITH-VERTEX-LOCAL-GUARANTEES *on complete graphs.*

Disagreement Minimization on Complete (Unweighted) Bipartite Graphs

A 10-approximation algorithm. Puleo and Milenkovic [2018] address the (complete) bipartite-graph with an approach similar to the one they adopt for complete graphs (described above). Specifically, they devise an algorithm to round a fractional clustering of a complete bipartite graph that, under Assumption 1, is proved to achieve a constant-factor error bound on every vertex of one of the two partite sets of the input graph. Formally, given a complete signed bipartite graph $G = (V_1, V_2, \mathcal{L})$—where V_1 and V_2 are the two partite sets of vertices, and $\mathcal{L} : V_1 \times V_2 \to \{-1, +1\}$ is a function labeling an edge as either positive or negative—Puleo and Milenkovic [2018]'s algorithm transforms any fractional clustering $\vec{c}$ of G into a hard clustering ℓ such that v-err$(\ell)_u \leq W$v-err$(\vec{c})_u$, for all $u \in V_1$ and some constant $W > 1$. Note that Puleo and Milenkovic [2018]'s algorithm does not guarantee any upper bound on v-err$(\vec{c})_u$ for $u \in V_2$. In this regard, Puleo and Milenkovic [2018] point out that, as the algorithm treats the sides V_1 and V_2 asymmetrically, it is difficult to control the per-vertex error at V_2. Nevertheless, an error bound for the vertices in V_1 suffices for some applications, e.g., in recommender systems, where vertices in V_1 correspond to users, while vertices in V_2 correspond to objects to be recommended, and quality of service only needs to be guaranteed for users, not for objects.

The pseudocode of Puleo and Milenkovic [2018]'s rounding algorithm for bipartite graphs is reported as Algorithm 3.12. It is a slight variant of the aforementioned rounding algorithm devised for complete graphs (Algorithm 3.9). Puleo and Milenkovic [2018] point out that the analysis of Algorithm 3.12 is simpler than Algorithm 3.9, as the focus on errors only at V_1 eliminates the need for the "bad pivots" argument used for the analysis of Algorithm 3.9. This also leads to a smaller value of the approximation constant W, which is now (roughly) equal to 10 (instead of 48). The ultimate approximation guarantee of Algorithm 3.12 is as follows.

Theorem 3.14 [Puleo and Milenkovic, 2018].   *Under Assumption 1, and given a fractional clustering $\bar{c}$ that is a solution of the corresponding convex-programming relaxation of the* MIN-DISAGREE-WITH-VERTEX-LOCAL-GUARANTEES *problem instance, Algorithm 3.12 achieves a* $\max\{c_1, c_2, c_3\}$-*approximation factor for* MIN-DISAGREE-WITH-VERTEX-LOCAL-GUARANTEES *on complete bipartite graphs where disagreements are measured with respect to only one of the two partite vertex sets, with*

$$c_1 = \frac{1}{1 - 2\alpha} + \frac{1}{k_1 \alpha - \gamma},$$

$$c_2 = \max\left\{\frac{1}{(1 - k_1)\alpha}, \frac{1}{\gamma}, \frac{2}{\alpha}\right\},$$

$$c_3 = \max\left\{\frac{1}{1 - 2\alpha}, \frac{2}{\alpha}\right\},$$

and k_1 is a constant, with $\frac{\gamma}{\alpha} < k_1 \leq 1$.

Numerically searching for the optimal parameter values leads to an approximation ratio of 10, with the following values:

$$\alpha = 0.377, \qquad \gamma = 0.102, \qquad k_1 = 0.730.$$

A 7-approximation algorithm.   Similarly to the complete-graph case, Charikar et al. [2017] improve Puleo and Milenkovic [2018]'s result on bipartite graphs as well. Specifically, Charikar et al. [2017] propose an algorithm—whose pseudocode is shown as Algorithm 3.13—that is a variant of the rounding algorithm they devised for complete graphs (Algorithm 3.10) and is shown to achieve an approximation factor of 7, under the same assumptions as the ones originally made by Puleo and Milenkovic [2018].

Algorithm 3.13 Bipartite-Rounding-Min-Disagree-with-Vertex-Local-Guarantees [Charikar et al., 2017]

Input: A complete unweighted signed bipartite graph $G = (V_1, V_2, \mathcal{L})$, a fractional clustering $\vec{c}$ of G

Output: A hard clustering ℓ of G

1: $S \leftarrow V$
2: **while** $S \cap V_1 \neq \emptyset$ **do**
3:    For each $v \in V_1 \cap S$, let $T_v = \{w \in S : \vec{c}_{vw} < \frac{3}{7}\}$ and $T_v^* = \{w \in V_2 \cap S : \vec{c}_{vw} < \frac{1}{7}\}$
4:    Choose a pivot vertex $u \in V_1 \cap S$ that maximizes $|T_u^*|$
5:    Add cluster T_u to the output ℓ clustering
6:    $S \leftarrow S \setminus T_u$
7: **end while**
8: Add every vertex in S (if any) as a singleton cluster to the output ℓ clustering

---

Algorithm 3.14 Bipartite-Rounding-Min-Disagree-with-Vertex-Local-Guarantees [Kalhan et al., 2019]

Input: A complete unweighted signed bipartite graph $G = (V_1, V_2, \mathcal{L})$, a fractional clustering $\vec{c}$ of G

Output: A hard clustering ℓ of G

1: $S \leftarrow V$
2: **while** $S \cap V_1 \neq \emptyset$ **do**
3:    For each $v \in V_1 \cap S$, let $T_v = \{w \in S : \vec{c}_{vw} \leq \frac{2}{5}\}$ and $T_v^* = \{w \in V_2 \cap S : \vec{c}_{vw} \leq \frac{1}{5}\}$
4:    Choose a pivot vertex $u \in V_1 \cap S$ that maximizes $\sum_{w \in T_u^*} \left(\frac{1}{5} - \vec{c}_{uw}\right)$
5:    Add cluster T_u to the output ℓ clustering
6:    $S \leftarrow S \setminus T_u$
7: **end while**
8: Add every vertex in S (if any) as a singleton cluster to the output ℓ clustering

---

Theorem 3.15   [Charikar et al., 2017].   *Under Assumption 1, and given a fractional clustering $\vec{c}$ that is a solution of the corresponding convex-programming relaxation of the* Min-Disagree-with-Vertex-Local-Guarantees *problem instance, Algorithm 3.13 achieves a 7-approximation factor for* Min-Disagree-with-Vertex-Local-Guarantees *on complete bipartite graphs where disagreements are measured with respect to only one of the two partite vertex sets.*

A 5-approximation algorithm.   For aggregation functions f corresponding to some ℓ_p-norm, Kalhan et al. [2019] further improve the approximation guarantee, getting to a 5-factor. Kalhan et al. [2019]'s algorithm is outlined as Algorithm 3.14, and is a slight variant of the 5-approximation algorithm Kalhan et al. [2019] devise for the complete-graph case (Algorithm 3.11).

Theorem 3.16   [Kalhan et al., 2019].   *For an aggregation function f corresponding to an ℓ_p-norm, for some $p \geq 1$, and given a fractional clustering $\vec{c}$ that is a solution of the corresponding convex-programming relaxation of the* Min-Disagree-with-Vertex-Local-Guarantees *problem instance, Algorithm 3.14 achieves a 5-approximation factor for* Min-Disagree-with-Vertex-Local-Guarantees *on complete bipartite graphs where disagreements are measured with respect to only one of the two partite vertex sets.*

Disagreement Minimization on General Weighted Graphs

Results for the ℓ_∞-norm.   When the aggregation function f corresponds to the ℓ_∞-norm, Min-Disagree-with-Vertex-Local-Guarantees corresponds to minimizing the maximum local disagreement (i.e., the maximum total weight of misclustered edges incident to any vertex). For this reason, Min-Disagree-with-Vertex-Local-Guarantees under the ℓ_∞-norm is also termed MinMax-Disagree-with-Vertex-Local-Guarantees.

The first result on general graphs is due to Charikar et al. [2017], who show that MinMax-Disagree-with-Vertex-Local-Guarantees admits an $\mathcal{O}(\sqrt{n})$-approximation. To this purpose, Charikar et al. [2017] point out that the nonlinear nature of correlation clustering with local objectives makes it much harder to approximate than the counterpart with classic global objectives. A major difficulty in this regard is that, while the natural LP relaxation for the classic Min-Disagree problem has a bounded integrality gap of $\mathcal{O}(\log n)$ [Charikar et al., 2005, Demaine et al., 2006] (see Section 1.5.3), both the natural LP and SDP relaxations of MinMax-Disagree-with-Vertex-Local-Guarantees have a much larger integrality gap of $\frac{n}{2}$ (as shown by Charikar et al. [2017] themselves).

Another critical point is that randomization is inherently difficult for local objectives, since a bound on the expected weight of misclassified edges incident on any vertex does not translate to a bound on the maximum of this quantity over all vertices (and, conversely, the same holds for the expected weight of correctly classified edges incident on any vertex, which does not translate to a bound on the minimum of this quantity over all the vertices).

To overcome these difficulties, Charikar et al. [2017] devise an algorithm that uses a combination of the LP lower bound and a combinatorial bound. The combinatorial bound is obtained with a (deterministic) greedy method that resembles the ones used for Min-Disagree-with-Vertex-Local-Guarantees on complete (bipartite) graphs (Algorithms 3.9—3.14), where a certain pivot vertex is iteratively picked, and a cluster is built by taking a sphere of a fixed and predefined radius around that pivot.

Although both the LP and combinatorial bounds on their own are not tight, their combination is anyway proved to achieve the aforementioned $\mathcal{O}(\sqrt{n})$-approximation. All details of the algorithm and its theoretical analysis can be found in Charikar et al. [2017].

Results for ℓ_p-norms, $p \geq 1$.   Kalhan et al. [2019] generalize the result of Charikar et al. [2017] by showing that Min-Disagree-with-Vertex-Local-Guarantees admits an

---

Algorithm 3.15 Non-Oblivious-Local-Search [Charikar et al., 2017]

Input: A weighted signed graph $G = (V, E, w, \mathcal{L})$, a constant $\varepsilon > 0$
Output: A clustering ℓ of G
1: $c^* \leftarrow \min_{u \in V} \sum_{v \in V : (u,v) \in E} w(u, v);$ $i \leftarrow 0;$ pick an arbitrary $S_0 \subseteq V$
2: let ℓ be composed of clusters S_0 and $V \setminus S_0$
3: **while** $\exists u \in V$ s.t. v-agree$(\ell)_u < \left(\frac{1}{2} - \varepsilon\right) c^*$ **do**
4:     move u to the other side of the cut S_i, denote the resulting cut by S_{i+1}, and update ℓ accordingly
5:     $i \leftarrow i + 1$
6: **end while**

$\mathcal{O}(n^{\frac{1}{2} - \frac{1}{2p}} \log^{\frac{1}{2} + \frac{1}{2p}} n)$-approximation for aggregation function f corresponding to some ℓ_p-norm.

When $p = \infty$, Kalhan et al. [2019]'s result matches the guarantee provided by Charikar et al. [2017] (up to logarithmic factors).

When $p = 1$, Min-Disagree-with-Vertex-Local-Guarantees boils down to the classic Min-Disagree. In this case, Kalhan et al. [2019]'s result is consistent with the best known approximation guarantee of $\mathcal{O}(\log n)$ for Min-Disagree on general graphs (see Section 1.5.3).

The $\mathcal{O}(n^{\frac{1}{2} - \frac{1}{2p}} \log^{\frac{1}{2} + \frac{1}{2p}} n)$-approximation algorithm devised by Kalhan et al. [2019] is a randomized algorithm that exploits a convex relaxation of correlation clustering (which is fairly standard, similar to the ones used in, e.g., Charikar et al. [2005], Demaine et al. [2006], Garg et al. [1996]), along with a technique of padded metric space decompositions, which aims at partitioning an arbitrary metric space into pieces of small diameter (used in, e.g., Bartal [1996], Fakcharoenphol et al. [2004], Gupta et al. [2003], Rao [1999]). All details and theoretical analyses can be found in Kalhan et al. [2019].

Agreement Maximization on General Weighted Graphs

For what concerns the maximization version, Charikar et al. [2017] observe that Max-Agree-with-Vertex-Local-Guarantees is closely related to the computation of local optima for Max-Cut, and the computation of pure Nash equilibria in cut and party affiliation games [Balcan et al., 2009, Bhalgat et al., 2010, Christodoulou et al., 2012, Fabrikant et al., 2004, Schäffer and Yannakakis, 1991] (a well-studied special class of potential games [Monderer and Shapley, 1996]).

In the setting of party affiliation games, each vertex of the input graph G is a player that can choose one of two sides of a cut. The player's payoff is the total weight of edges incident on it that are classified correctly. It is well known that such games admit a pure Nash equilibrium via the *best response dynamics* (also known as *Nash dynamics*), and that each such a pure Nash equilibrium is a $\frac{1}{2}$-approximation for MAX-AGREE-WITH-VERTEX-LOCAL-GUARANTEES.

Unfortunately, the computation of a pure Nash equilibria in cut and party affiliation games is in general **PLS**-complete [Johnson et al., 1988], and thus it is widely believed no polynomial-time algorithm exists for solving this problem. Nonetheless, one can apply the algorithm of Bhalgat et al. [2010] for finding an approximate pure Nash equilibrium and obtain a $\frac{1}{4+\varepsilon}$-approximation for MAX-AGREE-WITH-VERTEX-LOCAL-GUARANTEES on general weighted graphs (for any constant $\varepsilon > 0$).

Charikar et al. [2017] also focus on the special case where the aggregation function f corresponds to the ℓ_∞-norm. This variant of the problem, which, as said above, is alternatively referred to as MAXMIN-AGREE-WITH-VERTEX-LOCAL-GUARANTEES, corresponds to maximizing the minimum local agreement, i.e., the minimum total weight of correctly clustered edges incident on any vertex. The main result of Charikar et al. [2017] in this regard is that the approximation of MAXMIN-AGREE-WITH-VERTEX-LOCAL-GUARANTEES on general weighted graphs can be improved to $\frac{1}{2+\varepsilon}$. To this purpose, Charikar et al. [2017] notice that the natural local-search algorithm for MAXMIN-AGREE-WITH-VERTEX-LOCAL-GUARANTEES can be defined similarly to that of MAX-CUT: it maintains a single cut $S \subseteq V$; a vertex u moves to the other side of the cut if the move increases the total weight of correctly clustered edges incident on u. The algorithm terminates in a local optimum that is a $\frac{1}{2}$-approximation for MAXMIN-AGREE-WITH-VERTEX-LOCAL-GUARANTEES. Unfortunately, it is known that such a local-search algorithm can take exponential time, even for MAX-CUT.

When considering MAX-CUT, this can be remedied by altering the local-search step as follows: a vertex u moves to the other side of the cut S if the move increases the total weight of edges crossing S by a multiplicative factor of at least $(1 + \varepsilon)$ (for some $\varepsilon > 0$). But this approach *fails* for the computation of (approximate) pure Nash equilibria in party affiliation games, as well as for MAXMIN-AGREE-WITH-VERTEX-LOCAL-GUARANTEES. The reason is that both of these problems have *local* requirements from vertices, as opposed to the *global* objective of MAX-CUT. As mentioned above, a viable option is to employ Bhalgat et al. [2010]'s $\frac{1}{4+\varepsilon}$-approximation algorithm.

Charikar et al. [2017] improve such an approximation (to $\frac{1}{2+\varepsilon}$) by providing an alternative yet direct local-search approach that circumvents the need for computing approximate pure Nash equilibria in party affiliation games. Specifically, Charikar et al. [2017] define a *non-oblivious* local search that is executed with *altered* edge weights, i.e., with edge weights changed in such a way that: (i) any local optimum is a $\frac{1}{2+\varepsilon}$-approximation, and (ii) the local search performs at most $\mathcal{O}(\frac{n}{\varepsilon})$ iterations. The pseudocode of such a non-oblivious local-search algorithm is reported in Algorithm 3.15, while its approximation guarantee is stated next.

Theorem 3.17   [Charikar et al., 2017].   *For any $\varepsilon > 0$, Algorithm 3.15 is a $\frac{1}{2+\varepsilon}$-approximation algorithm for MAXMIN-AGREE-WITH-VERTEX-LOCAL-GUARANTEES on general weighted graphs, and runs in* poly $\left(n, \frac{1}{\varepsilon}\right)$.

3.2.2 CLUSTER-WISE FORMULATION

The cluster-wise variant of correlation clustering with local objectives has received less attention than the vertex-wise counterpart. Indeed, no hardness results are known, while, in terms of approximation, results have been derived only for the MinMax-Disagree-with-Cluster-Local-Guarantees variant of the problem, i.e., for the special case of Min-Disagree-with-Cluster-Local-Guarantees where the aggregation function f corresponds to the ℓ_∞-norm.

The first approximation result for MinMax-Disagree-with-Cluster-Local-Guarantees is due to Ahmadi et al. [2019], who provides an $\mathcal{O}(\log n)$-approximation algorithm for general weighted graphs. Ahmadi et al. [2019]'s algorithm considers a minimax variant of the Min-Multicut problem, and properly adapts the approximation-preserving reduction between Min-Multicut and Min-Disagree devised in Demaine et al. [2006]. Ahmadi et al. [2019] also show that such an $\mathcal{O}(\log n)$-approximation improves to $\mathcal{O}(r^2)$ on graphs that exclude $K_{r,r}$ minors, and to a constant factor of 14 on complete (unweighted) graphs. The result of Ahmadi et al. [2019] on general weighted graphs is improved by Kalhan et al. [2019], who devise a $(2 + \varepsilon)$-approximation algorithm, for any $\varepsilon = 1/\text{poly}(n)$. The idea of Kalhan et al. [2019]'s algorithm consists in iteratively solving—via LP relaxation—a subproblem that, for a vertex $u \in V$, finds a subset $S \subseteq V$ containing u such that the local cost of MinMax-Disagree-with-Cluster-Local-Guarantees in correspondence of u is minimized.

3.3 CORRELATION CLUSTERING WITH OUTLIERS

So far, we have assumed that all vertices in the input graph need to be clustered. What happens if an *unknown* subset of vertices are outliers and should not be placed into any cluster? Aboud [2008] was the first to consider this version of the problem in his master thesis, inspired by the prize-collecting problems of network design. Aboud [2008]'s formulation, dubbed "correlation clustering with penalties" assumes that vertices may be discarded from the clustering at a certain penalty cost. Formally, Aboud [2008]'s problem is as follows.

Problem 3.18 (Correlation-Clustering-with-Outliers) Given a set V of objects, a pairwise similarity function $s : \binom{V}{2} \to \{0, 1\}$, and a penalty function $p : V \to \mathbb{R}$ find a set $D \subseteq V$ of outliers and clustering $\ell : V \setminus D \to \mathbb{N}$ of the remaining vertices minimizing the following cost:

$$\text{cost}(D, \ell) = \sum_{x \in D} p(x) + \sum_{\substack{(x,y) \in \binom{V \setminus D}{2}, \\ \ell(x) = \ell(y)}} (1 - s(x, y)) + \sum_{\substack{(x,y) \in \binom{V \setminus D}{2}, \\ \ell(x) \neq \ell(y)}} s(x, y).$$

$$(3.13)$$

By setting arbitrarily large penalties $p_x = \infty$, one recovers the MIN-DISAGREE problem. Hence, Problem 3.18 is **NP**-hard. Aboud [2008] tackles this problem and introduces a modification of LP (1.6) which takes into account the penalties in the objective function, and proposes a primal-dual method which can be used to detect a set of outliers before running QwickCluster. Aboud [2008] shows that this method yields a 9-approximation algorithm for Problem 3.18. Aboud [2008] also analyzes the more general variant of Problem 3.18 with weighted graphs, and the analysis is extended to obtain a 17-approximation algorithm for correlation clustering with outliers in complete *weighted* graphs, assuming the weights satisfy the probability constraint.

In the special case that each penalty $p(x)$ is equal to 1, a 6-approximation algorithm for Problem 3.18 was given by Devvrit et al. [2019]. In fact, they consider a related problem, termed ROBUST-CORRELATION-CLUSTERING, in which a hard upper bound m on the number of outliers is taken as an additional input; the objective function remains the same as in correlation clustering. Hence, in the case where each penalty $p(x)$ is equal to 1, Problem 3.18 may be viewed as a Lagrangian relaxation of the ROBUST-CORRELATION-CLUSTERING problem. Unlike Problem 3.18, it is **NP**-hard to approximate ROBUST-CORRELATION-CLUSTERING to within any constant factor, as shown in Devvrit et al. [2019] by a reduction from MIN-MULTICUT. Hence, they examined bicriteria approximation algorithms; an (α, β) bicriteria approximation for ROBUST-CORRELATION-CLUSTERING is one where the solution cost is at most α times the optimal cost, and the number of outliers in the solution is at most βm. Devvrit et al. [2019] give a $(6, 6)$ approximation for complete graphs. The algorithm essentially does a simple LP-based pre-processing step to prune out a set of $O(m)$ outliers and then runs the well-established QwickCluster algorithm [Ailon et al., 2008a] (see Section 1.5). Devvrit et al. [2019] also design an efficient $(O(\log n), O(\log^2 n))$ bicriteria approximation algorithm on general graphs using padded decompositions of metric spaces, which are interpreted as roundings of the fractional LP solutions into an integral clustering.

The problem of correlation clustering with outliers has also been taken into account by Bonchi et al. [2019] in the task of *discovering polarized communities in signed networks*. Specifically, Bonchi et al. [2019] formulate this task as a variant of the MAX-AGREE[2] problem (introduced in Section 2.1) with outliers, and develop two spectral algorithms for it.

CHAPTER 4

Other Types of Graphs

This chapter describes variants of correlation clustering that have a special type of input, e.g., simple graphs with a specific structure, information-richer graphs, or graphs with noisy or uncertain information.

In particular, this chapter discusses correlation clustering on bipartite graphs (Section 4.1), edge-labeled graphs (Section 4.2), multilayer graphs (Section 4.3), vertex-labeled graphs (Section 4.4), hypergraphs (Section 4.5), and noisy and uncertain graphs (Section 4.6).

4.1 BIPARTITE GRAPHS

The task of clustering the vertices of a bipartite graph is common across many areas of machine learning. Applications include recommender systems [Vlachos et al., 2014], where analyzing the structure of a large set of pairwise interactions (e.g., among users and products) enables useful predictions about future interactions, gene-expression-data analysis [Amit, 2004, Madeira and Oliveira, 2004], and graph-partitioning problems in data mining [Fern and Brodley, 2004, Zha et al., 2001].

Bipartite-Correlation-Clustering is a natural variant of correlation clustering on bipartite graphs, where the objective is to cover an input set of vertices with disjoint cliques (clusters), so as to minimize the symmetric difference with the given edge set over these vertices. Bipartite-Correlation-Clustering is a special case of correlation clustering on general graphs, thus algorithms for the latter [Charikar et al., 2003, Demaine et al., 2006, Swamy, 2004] can in principle be applied to Bipartite-Correlation-Clustering. However, major downsides of those algorithms include the fact that they do not leverage the structure of the bipartite graph, and that they rely on LP or SDP solvers, which scale poorly.

Following Ailon et al. [2011] and for sake of simplicity, we present here the unweighted case, although most of the results carry over straightforwardly to the weighted case.

> **Problem 4.1 (Bipartite-Correlation-Clustering)** Given a bipartite graph $G = (U, V, E)$ where U and V are the sets of left and right vertices and E is the set of edges, i.e., a function $E : U \times V \to \{0, 1\}$, the goal is to cluster the elements of $U \cup V$ in such a way that, to the best possible extent, the resulting clusters are disjoint bi-cliques. Assuming that cluster identifiers are

represented by natural numbers, a clustering ℓ is a function $\ell : U \cup V \to \mathbb{N}$. Bipartite-Correlation-Clustering aims at finding a clustering ℓ that minimizes the following cost:

$$\sum_{\substack{(u,v)\in U\times V, \\ \ell(x)=\ell(y)}} 1 - E(u,v) \sum_{\substack{(x,y)\in U\times V, \\ \ell(x)\neq\ell(y)}} E(u,v). \tag{4.1}$$

Amit [2004] was the first to address Bipartite-Correlation-Clustering directly. She proved its **NP**-hardness and gave a constant 11-approximation algorithm based on rounding a linear programming in the spirit of Charikar et al. [2005]'s algorithm for Min-Disagree.

Guo et al. [2008] provides a 4-approximation algorithm for Bipartite-Correlation-Clustering, which was later on shown to be incorrect by Ailon et al. [2011]. However, the attempt of using ideas from Ailon et al. [2008a] in Guo et al. [2008]'s algorithm inspired Ailon et al. [2011, 2012a] who devised an algorithm that achieves a 4-approximation. This algorithm of Ailon et al. [2011, 2012a], which provides the best guarantee currently known, is presented in Section 4.1.1.

Chawla et al. [2015] deal with (among others) the problem of correlation clustering on complete k-partite graphs, and devise a 3-approximation algorithm that is valid for any k, and, as such, valid for the bipartite ($k = 2$) case as well.

Asteris et al. [2016] study the version of Bipartite-Correlation-Clustering, where the number k of cluster is given as input and the goal is to maximize agreement. In particular, given a bipartite graph $G = (U, V, E)$, a parameter k, and any constant accuracy parameter $\delta \in (0, 1)$, Asteris et al. [2016] devise an algorithm that computes a clustering of $U \cup V$ into at most k clusters and achieves a number of agreements that lies within a $(1 - \delta)$-factor from the optimum. It runs in time exponential in k and δ^{-1}, but linear in the size of G.

Moreover, for the unconstrained (i.e., with no k given) Bipartite-Correlation-Clustering setting, the optimal number of clusters may be anywhere from 1 to $|U| + |V|$. Asteris et al. [2016] show that if one is willing to settle for a $(1 - \delta)$-approximation of the max-agreement objective, it suffices to use at most $O(\delta^{-1})$ clusters, regardless of the size of G. In turn, under an appropriate configuration, the algorithm by Asteris et al. [2016] for the constrained version of Bipartite-Correlation-Clustering yields an EPTAS[1] for the unconstrained problem.

---

[1]*Efficient Polynomial Time Approximation Scheme* (EPTAS) refers to an algorithm that approximates the solution of an optimization problem within a multiplicative $(1 - \epsilon)$-factor, for any constant $\epsilon \in (0, 1)$, and has complexity that scales arbitrarily in $1/\epsilon$, but as a constant order polynomial (independent of ϵ) in the input size n. EPTAS is more efficient than a Polynomial Time Approximation Scheme (PTAS); for example, a running time of $O(n^{1/\epsilon})$ is considered a PTAS, but not an EPTAS.

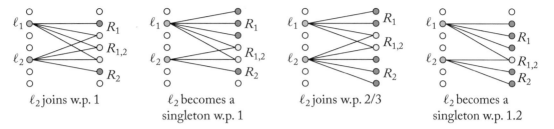

Figure 4.1: Pivot-BiCluster algorithm: four example cases in which ℓ_2 either joins the cluster created by ℓ_1 or becomes a singleton. In the two right-most examples, with the remaining probability nothing is decided about ℓ_2.

Veldt et al. [2020] introduce a general parameterized framework that allows for capturing different graph-clustering objectives—including BIPARTITE-CORRELATION-CLUSTERING—by properly setting some resolution parameters.

4.1.1   PIVOT-BICLUSTER ALGORITHM

We next describe the 4-approximation randomized algorithm by Ailon et al. [2011, 2012a] for the min-disagreement version of BIPARTITE-CORRELATION-CLUSTERING (i.e., Problem 4.1). The algorithm, termed Pivot-BiCluster, is a sequential one. In every cycle it creates one cluster and possibly many singletons, all of which are removed from the graph before continuing to the next iteration. Slightly abusing notation, in the description of the algorithm' we denote by $N(\ell)$ all the neighbors of $\ell \in U$ which have not been removed from the graph yet.

Every such cycle performs two phases. In the first phase, Pivot-BiCluster picks a node on the left side uniformly at random, ℓ_1, and forms a new cluster $C = \{\ell_1\} \cup N(\ell_1)$. This will be referred to as the ℓ_1-phase and ℓ_1 will be referred to as the left center of the cluster. In the second phase, denoted as the ℓ_2-sub-phase corresponding to the ℓ_1-phase, the algorithm iterates over all the other remaining left nodes, ℓ_2, and decides either to (1) append them to C, (2) turn them into singletons, or (3) do nothing. We now explain how to make this decision. Let $R_1 = N(\ell_1) \setminus N(\ell_2)$, $R_2 = N(\ell_2) \setminus N(\ell_1)$ and $R_{1,2} = N(\ell_1) \cap N(\ell_2)$. With probability $\min\{\frac{|R_{1,2}|}{|R_2|}, 1\}$ do one of two things: (1) if $|R_{1,2}| \geq |R_1|$ append ℓ_2 to C, and otherwise (2) (if $|R_{1,2}| < |R_1|$), turn ℓ_2 into a singleton. In the remaining probability, (3) do nothing for ℓ_2, leaving it in the graph for future iterations. Examples for cases the algorithm encounters for different ratios of R_1, $R_{1,2}$, and R_2 are given in Figure 4.1.

The main result of Ailon et al. [2011] is the analysis of the Pivot-BiCluster algorithm, which is inspired by that for the QwickCluster algorithm [Ailon et al., 2008a] for standard correlation clustering (see Section 1.5).

Theorem 4.2   [Ailon et al., 2011].   *The Pivot-BiCluster algorithm returns a solution to* BIPARTITE-CORRELATION-CLUSTERING *with expected cost at most four times that of the optimal solution.*

Due to the complexity of the approximation analysis, we do not report it here and instead refer the interested reader to Ailon et al. [2011, 2012a].

4.2   EDGE-LABELED GRAPHS

A common trait underlying most clustering paradigms is the existence of a *real-valued proximity function* $f(\cdot, \cdot)$ representing the similarity/distance between pairs of objects. Bonchi et al. [2012, 2015] consider a clustering setting where the relation among objects is represented by a relation type, that is a label from a finite set of possible labels L, while a special label $l_0 \notin L$ is used to denote that the objects have no relation. In other words, they consider the case where the range of the proximity function $f(\cdot, \cdot)$ is *categorical*, instead of *numerical*. This setting has a natural graph interpretation: the input can be viewed as an *edge-labeled graph* $G = (V, E, L, l_0, \ell)$, where the set V of vertices corresponds to the objects to be clustered, the set E of edges comprises all unordered pairs within V having some relation (i.e., whose relation is represented by a label other than the l_0 label), and the function ℓ assigns to each edge in E a label from L.

The study of edge-labeled graphs is motivated by many real-world applications and has received considerable attention in the data-mining literature in the last years [Bonchi et al., 2014b, Fan et al., 2011, Jin et al., 2010, Khan et al., 2015, 2018]. As an example, social networks are commonly represented as graphs, where the vertices represent individuals and the edges model relationships among these individuals: these relationships can be of various types, e.g., colleagues, neighbors, schoolmates, teammates. Biologists study protein-protein interaction networks, where vertices represent proteins and edges represent interactions occurring when two or more proteins bind together to carry out their biological function. Those interactions can be of different types, e.g., physical association, direct interaction, co-localization, etc. In these networks, for instance, a cluster containing mainly edges labeled as co-localization, might represent a *protein complex*, i.e., a group of proteins that interact with each other at the same time and place, forming a single multi-molecular machine [Lin et al., 2007]. In bibliographic data, co-authorship networks represent collaborations among authors: in this case, the topic of the collaboration can be viewed as an edge label, and a cluster represents a topic-coherent community of researchers.

According to these application examples, an interesting goal in clustering edge-labeled graphs is to find a partition of the vertices of the graph such that the edges in each cluster have, as much as possible, the same label. An example is shown in Figure 4.2, where colors are used to indicate labels. Intuitively, a red edge (x, y) provides positive evidence that the vertices x and y should be clustered in such a way that the edges in the subgraph induced by that cluster are mostly red. Furthermore, in the case that most edges of a cluster are red, it is reasonable to label the whole cluster with the red color.

Note that a clustering algorithm for this problem should also deal with inconsistent evidence, as a red edge (x, y) provides evidence for the vertex x to participate in a cluster with red edges, while a green edge (x, z) provides contradicting evidence for the vertex x to partic-

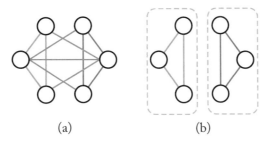

(a)   (b)

Figure 4.2: An example of chromatic clustering: (a) input graph and (b) the optimal solution for CHROMATIC-CORRELATION-CLUSTERING (Problem 4.3).

ipate in a cluster with red edges. Aggregating such an inconsistent information is resolved by optimizing an objective function that takes into account such designing principles. Given the natural representation of an edge label as its color, Bonchi et al. [2012, 2015] dub this problem CHROMATIC-CORRELATION-CLUSTERING, and formulate it as follows.

Problem 4.3 (Chromatic-Correlation-Clustering)   Given an edge-labeled graph $G = (V, E, L, l_0, \ell)$, where V is a set of vertices, $E \subseteq V_2$ is a set of edges (here $V_2 = V \times V$), L is a set of labels, $l_0 \notin L$ is a *special* label, and $\ell : V_2 \to L \cup \{l_0\}$ is a labeling function that assigns a label to each unordered pair of vertices in V such that $\ell(x, y) = l_0$ if and only if $(x, y) \notin E$, find a clustering $C : V \to \mathbb{N}$ and a cluster-labeling function $\ell : C[V] \to L$ so as to minimize the cost

$$\text{cost}(G, C, \ell) = \sum_{\substack{(x,y)\in V_2, \\ C(x)=C(y)}} (1 - \mathbb{1}[\ell(x, y) = \ell(C(x))]) + \sum_{\substack{(x,y)\in V_2, \\ C(x)\neq C(y)}} (1 - \mathbb{1}[\ell(x, y) = l_0]), \quad (4.2)$$

where $\mathbb{1}[\cdot]$ denotes the indicator function.

Equation (4.2) is composed of two terms, representing intra- and inter-cluster costs, respectively. In particular, the intra-cluster-cost term aims to measure the homogeneity of the labels on the intra-cluster edges: a pair of vertices (x, y) assigned to the same cluster pays a cost if and only if their relation type $\ell(x, y)$ is different from the predominant relation type of the cluster, as indicated by the ℓ function. On the other hand, as far as the inter-cluster cost, the objective function penalizes two vertices x and y unless they are adjacent (i.e., unless $\ell(x, y) \neq l_0$); instead, if x and y are adjacent, the objective function incurs a cost, regardless of the label $\ell(x, y)$ on the shared edge.

Example 4.4   Consider the problem instance in Figure 4.2a, along with the solution in Figure 4.2b, where the two clusters depicted are labeled with the green and blue label, respectively. The intra-cluster cost of the solution is equal to the number of intra-cluster edges that have a label different from the one assigned to the corresponding cluster. As both the clusters are monochromatic cliques, the resulting overall intra-cluster cost is zero. The overall cost of the solution is thus equal to the inter-cluster cost only, which corresponds to the number of edges between vertices in different clusters and is, therefore, equal to 5.

It is easy to observe that, when $|L| = 1$, the CHROMATIC-CORRELATION-CLUSTERING problem corresponds to the standard (min-disagree) formulation of correlation clustering. Since CHROMATIC-CORRELATION-CLUSTERING is a generalization of MIN-DISAGREE, which is, as known, **NP**-hard [Bansal et al., 2004], one can easily conclude that CHROMATIC-CORRE-LATION-CLUSTERING is **NP**-hard as well. Based on this observation, it is natural to ask whether applying standard correlation-clustering algorithms, just ignoring the different colors, would be a good solution to the problem. The next example shows that such an approach is not guaranteed to produce a good solution.

Example 4.5   For the problem instance in Figure 4.2a, the optimal solution of the standard MIN-DISAGREE problem that ignores edge colors, would correspond to a single cluster containing all the six vertices, as, according to Equation (1.1), this solution has a (minimum) cost of 4 corresponding to the number of missing edges within the cluster. Conversely, this solution has a non-optimal cost 12 when evaluated in terms of the CHROMATIC-CORRELATION-CLUSTERING objective function, i.e., according to Equation (4.2). Instead, the optimum in this case corresponds to the cost-5 solution depicted in Figure 4.2b.

4.2.1   APPROXIMATION ALGORITHMS

The first algorithm proposed in Bonchi et al. [2012] is a randomized approximation algorithm inspired by the QwickCluster algorithm [Ailon et al., 2008a] for standard correlation clustering (see Section 1.5).

The pseudocode of the randomized approximation algorithm, here dubbed ChromaticQwickCluster, is reported as Algorithm 4.16. The main difference with the QwickCluster algorithm is that the edge labels are taken into account in order to build label-homogeneous clusters around the pivots. To this end, the pivot chosen at each iteration of ChromaticQwickCluster is an edge (x, y), rather than a single vertex. The pivot edge is used to build a cluster around it: beyond the vertices x and y, the cluster C being formed contains all other vertices z still in the graph for which the *triangle* (x, y, z) is monochromatic, that is, $\ell(x, y) = \ell(y, z) = \ell(x, y)$. Since the label $\ell(x, y)$ forms the basis for creating the cluster C, the cluster inherits this label. All vertices in C along with all their incident edges are removed from the graph, and the main cycle of the algorithm terminates when there are no edges anymore. Eventually, all the remaining vertices (if any) are made singleton clusters. It is easy to see that the ChromaticQwickCluster algorithm runs in $\mathcal{O}(|V| + |E|)$ time.

Algorithm 4.16 ChromaticQwickCluster

Input: Edge-labeled graph $G = (V, E, L, l_0, \ell)$, where $\ell : V_2 \to L \cup \{l_0\}$
Output: Clustering $\mathcal{C} : V \to \mathbb{N}$; cluster labeling function $\ell : \mathcal{C}[V] \to L$

  $i \leftarrow 1$
  while $E \neq \emptyset$ **do**
    pick an edge $(x, y) \in E$ uniformly at random
    $C \leftarrow \{x, y\} \cup \{z \in V \mid \ell(x, z) = \ell(y, z) = \ell(x, y)\}$
    $\mathcal{C}(x) \leftarrow i$, for all $x \in C$
    $\ell(i) = \ell(x, y)$
    remove C from G: $V \leftarrow V \setminus C$, $E \leftarrow E \setminus \{(x, y) \in E \mid x \in C\}$
    $i \leftarrow i + 1$
  end while
  for all $x \in V$ **do**
    $\mathcal{C}(x) \leftarrow i$
    $\ell(i) \leftarrow$ a label from L
    $i \leftarrow i + 1$
  end for

The main result presented by Bonchi et al. [2012] is an approximation guarantee for the solutions output by ChromaticQwickCluster. In particular, Theorem 4.8, shows that the approximation guarantee of the algorithm depends on the number of *bad triplets* incident on a pair of vertices in the input graph. Even though the ChromaticQwickCluster algorithm is similar to the QwickCluster algorithm, the theoretical analysis of ChromaticQwickCluster is much more complicated and requires several additional and nontrivial arguments. We overview this analysis next.

Let a subset of three vertices $\{x, y, z\}$ be a *bad triplet* (B-triplet) if the induced triangle has at least two edges and is non-monochromatic, i.e., $|\{(x, y), (x, z), (y, z)\} \cap E| \geq 2$ and $|\{\ell(x, y), \ell(x, z), \ell(y, z)\}| > 1$. Let $\mathcal{T}$ denote the set of all B-triplets for an instance of Problem 4.3. Moreover, given a pair $(x, y) \in V_2$, let $\mathcal{T}_{xy} \subseteq \mathcal{T}$ denote the set of all B-triplets in $\mathcal{T}$ that contain both x and y, i.e., $\mathcal{T}_{xy} = \{t \in \mathcal{T} \mid x \in t, y \in t\}$, while $T_{max} = \max_{(x,y) \in V_2} |\mathcal{T}_{xy}|$ denotes the maximum number of B-triplets that contain a pair of vertices.

Now, consider an instance $G = (V, E, L, l_0, \ell)$ of Problem 4.3 and the output $\langle \mathcal{C}, c\ell \rangle$ of the ChromaticQwickCluster algorithm on G, where, we recall, $\mathcal{C}$ is the output clustering while $c\ell$ is the cluster labeling function. Rewrite the cost function in Equation (4.2) as the sum of the costs paid by a single pair $(x, y) \in V_2$. To this end, in order to simplify the notation, we hereinafter write the cost by omitting $\mathcal{C}$ and ℓ while keeping G only:

$$\text{cost}(G) = \sum_{(x,y) \in V_2} \text{cost}_{xy}(G), \tag{4.3}$$

where $\text{cost}_{xy}(G)$ denotes the aforementioned contribution of the pair (x, y) to the total cost.

The first result exploited by Bonchi et al. [2012] in their analysis is the following: a pair of vertices (x, y) pays a non-zero cost in a solution output by ChromaticQwickCluster *only if* such a pair belongs to at least one B-triplet of the input graph. This result is formally stated in the following lemma.

Lemma 4.6   [Bonchi et al., 2012]   *If* $\mathrm{cost}_{xy}(G) > 0$, *then* $\mathcal{T}_{xy} \neq \emptyset$.

Proof. According to the cost function defined in Equation (4.2), $\mathrm{cost}_{xy}(G) > 0$ if and only if either (1) x and y are adjacent and the edge (x, y) is split (i.e., x and y are put in different clusters), or (2) x and y are placed in the same cluster C while $\ell(x, y)$ is not equal to the label of C. We analyze each case next.

(1) According to the outline of ChromaticQwickCluster, (x, y) is split when, at some iteration i, it happens that x is put into cluster C, while y is not, or vice versa. Assuming that the vertex chosen to belong to C is x (an analogous reasoning holds considering y as belonging to C), we have two subcases:

  (a) An edge (x, z) is chosen as pivot, with $z \neq y$. The fact that the edge (x, y) is split implies one of the following: either $\ell(x, y) \neq \ell(x, z)$ or $\ell(y, z) \neq \ell(x, z)$; each of the cases further implies that the triangle $\{x, y, z\}$ is a B-triplet.

  (b) The edge chosen as pivot is (z, w), with $\{z, w\} \cap \{x, y\} = \emptyset$. In this case, to have (i) (x, y) split and (ii) x put in the cluster being formed while y does not, it must be verified that either $\ell(z, y) \neq \ell(z, w)$ or $\ell(w, y) \neq \ell(z, w)$. Each of these cases implies that there exists a B-triplet containing x and y (i.e., either $\{x, y, z\}$ or $\{x, y, w\}$).

(2) In this case, as both x and y are chosen as being part of the current cluster C, it must hold that the pivot chosen is (z, w), with $\{z, w\} \cap \{x, y\} = \emptyset$, and both $\{x, z, w\}$ and $\{x, y, w\}$ are monochromatic triangles. This fact, along with the hypothesis $\ell(x, y) \neq \ell(z, w)$, implies that both $\{x, y, z\}$ and $\{x, y, w\}$ are B-triplets.

The above reasoning shows that all situations where the pair (x, y) pays a non-zero cost imply the existence of at least one B-triplet that contains x and y. The lemma follows.   □

A direct implication of Lemma 4.6 is that one can express the cost of the solution of the ChromaticQwickCluster algorithm in terms of B-triplets only. Indeed, the lemma guarantees that vertex pairs that are not contained in a B-triplet pay no cost; thus, all non-B-triplets can safely be discarded. For every $t \in \mathcal{T}$, let α_t denote the cost paid by t in a ChromaticQwickCluster solution. Note that α_t can be less than 3, because not all the three vertex pairs of t are charged against t, as a vertex pair may participate in more than one B-triplet of the input graph, but the pair's contribution to the overall cost is at most one. As a consequence, the $\mathrm{cost}(G)$ can be expressed as:

$$\mathrm{cost}(G) = \sum_{t \in \mathcal{T}} \alpha_t.$$

Let us now focus on the cost of an *optimal* solution to the input problem instance G. Let $\text{cost}^*(G)$ denote such an optimal (i.e., minimum) cost. As shown in the next lemma, we can obtain a lower bound of $\text{cost}^*(G)$ by using α_t.

Lemma 4.7 **[Bonchi et al., 2012]**   *The cost of the optimal solution on graph G has the following bound*

$$\text{cost}^*(G) \geq \frac{1}{3T_{max}} \sum_{t \in \mathcal{T}} \alpha_t.$$

Proof. We start by observing that a B-triplet incurs a non-zero cost in every solution, and thus in the optimal solution as well. Here, by "a non-zero cost incurred by a B-triplet" we mean a non-zero cost paid by at least one of the vertex pairs in that B-triplet. Then, a lower bound on the cost of the optimal solution $\text{cost}^*(G)$ can be obtained by counting the number of disjoint B-triplets in the input. This lower bound can alternatively be expressed by considering the whole set of (not necessarily edge-disjoint) B-triplets in the input, by restating the following result of Ailon et al. [2008a]: denoting by $\{\beta_t \mid t \in \mathcal{T}\}$ any assignment of nonnegative weights to the B-triplets in $\mathcal{T}$ satisfying $\sum_{t \in \mathcal{T}_{xy}} \beta_t \leq 1$ for all $(x, y) \in V_2$, it holds that $\text{cost}^*(G) \geq \sum_{t \in \mathcal{T}} \beta_t$.

Now, we note that, as $\alpha_t \leq 3$, then, for every vertex pair (x, y) it holds that $\sum_{t \in \mathcal{T}_{xy}} \alpha_t \leq 3|\mathcal{T}_{xy}| \leq 3T_{max}$; and thus $\sum_{t \in \mathcal{T}_{xy}} \frac{\alpha_t}{3T_{max}} \leq 1$, for all $(x, y) \in V_2$. This implies that setting $\beta_t = \frac{\alpha_t}{3T_{max}}$ suffices to satisfy the condition $\sum_{t \in \mathcal{T}_{xy}} \beta_t \leq 1$ for all $(x, y) \in V_2$, and thus, the result by Ailon et al. applies:

$$\text{cost}^*(G) \geq \sum_{t \in \mathcal{T}} \beta_t = \frac{1}{3T_{max}} \sum_{t \in \mathcal{T}} \alpha_t.$$

□

The final approximation factor of the ChromaticQwickCluster algorithm can now be easily derived by combining Lemmas 4.6 and 4.7. Such a result is formally stated in the following theorem.

Theorem 4.8 **[Bonchi et al., 2012].**   *The approximation ratio of the ChromaticQwickCluster algorithm on input G is*

$$\frac{\text{cost}(G)}{\text{cost}^*(G)} \leq 3T_{max}.$$

Proof. Immediate from Lemmas 4.6 and 4.7:

$$\frac{\text{cost}(G)}{\text{cost}^*(G)} \leq \frac{\sum_{t \in \mathcal{T}} \alpha_t}{\frac{1}{3T_{max}} \sum_{t \in \mathcal{T}} \alpha_t} = 3T_{max}.$$

□

Theorem 4.8 shows that the approximation factor of the ChromaticQwickCluster algorithm is bounded by the maximum number T_{max} of B-triplets that contain a specific pair of vertices. The result quantifies the quality of the performance of the algorithm as a property of the input graph. As an example, Bonchi et al. [2012] show that the algorithm provides a constant-factor approximation for bounded-degree graphs.

Constant-Factor Approximation Algorithms

Anava et al. [2015] improve the results of Bonchi et al. [2012] devising constant-factor approximation algorithms. Their first contribution is a framework that builds on an observation that reduces an underlying instance of CHROMATIC-CORRELATION-CLUSTERING to that of standard correlation clustering, while losing only a small constant factor in the approximation. One can then employ any of the constant-factor approximation algorithms developed for correlation clustering and attain a constant-factor approximation. For example, one can utilize the fast randomized QwickCluster algorithm of Ailon et al. [2008a] to obtain a linear-time randomized 11-approximation algorithm.

Anava et al. [2015]'s algorithmic framework consists of two main steps:

1. The algorithm initially modifies the input graph G. Specifically, the algorithm processes the vertices of the graph one by one, and associates a positive color with each one of them. The color $c_v \in L$ that is associated with a node v is the positive color that appears the most among the edges that are incident to v in G, breaking ties arbitrarily. Then, the algorithm modifies the color of all the edges incident to each node v that are different from c_v to the special negative color λ. At the end of this step, each edge of positive color $c \in L$ in the modified graph G' must be incident to two nodes whose associated color was c. This implies that if we only focus on positive edges and neglect all the negative λ-edges from G', then each connected component has exactly the same color, henceforth referred to as the color of the connected component.

2. The algorithm continues by executing any fast algorithm for the standard correlation-clustering problem on each of the previously-mentioned connected components of the modified graph G'. All the clusters in each resulting clustering are given the positive color of the underlying connected component. Note that the ground sets of vertices of the resulting clusterings are pairwise disjoint, and therefore, the algorithm proceeds by uniting all those clustering to one clustering, which is then returned as the output of the algorithm.

Anava et al. [2015] focus the analysis of the above framework where the algorithm utilized in its second step is the randomized pivot-based QwickCluster algorithm of Ailon et al. [2008a] (that we discussed in Section 1.5): pick a pivot vertex uniformly at random, create a cluster that consists of this pivot and all the adjacent vertices that are connected to it using a positive edge, remove those vertices from the graph, and repeat as long as there are still vertices that are not clustered. It is worth noting that the properties of the pivot-based algorithm enable its direct

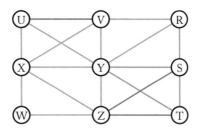

Figure 4.3: An example of an edge-labeled graph.

application to the entire modified graph, while treating all positive color edges identically, without the need to identify each of its connected components first. We refer to the paper by Anava et al. [2015] for the detailed approximation analysis of the proposed framework.

Anava et al. [2015] also develop a linear-programming-based algorithm, which applies deterministic rounding, and achieves an improved approximation ratio of 4. This algorithm builds on the approach of Charikar et al. [2003] for standard correlation clustering, augmenting it with additional insights and observations that enable its generalization to the chromatic setting. Although this algorithm cannot be considered to be practical, it extends the theoretical understanding of the Chromatic-Correlation-Clustering problem. In fact, this approach can be adjusted to the arbitrary graph setting, attaining a $O(\log |V|)$-approximation. This matches the best approximation known for the special case of correlation clustering on general graphs [Charikar et al., 2005, Demaine et al., 2006].

The result of Anava et al. [2015] is further improved by Klodt et al. [2021], who show that when simply run color-blind, the QwickCluster algorithm is a (linear-time) 3-approximation algorithm for Chromatic-Correlation-Clustering.

4.2.2   HEURISTICS

Besides the theoretical results presented in the previous section, both Bonchi et al. [2012] and Anava et al. [2015] propose more practical heuristic algorithms, that we discuss next.

The first heuristic algorithm proposed by Bonchi et al. [2012] is a variant of the ChromaticQwickCluster algorithm (Algorithm 4.16) that employs two heuristics in order to avoid some potential bad behavior of the algorithm, as exemplified next.

Example 4.9   Consider the graph in Figure 4.3: it has a fairly evident green cluster composed of all vertices in the graph but S and T (i.e, vertices {U,V,R,X,Y,W,Z}). However, as all the edges have the same probability of being selected as pivots, ChromaticQwickCluster might miss this green cluster, depending on which edge is selected first. For instance, suppose that the first pivot chosen is (Y,S). ChromaticQwickCluster forms the red cluster {Y,S,T} and removes it from the graph. Removing vertex Y makes the edge (X,Y) missing, which would have been a good pivot to build a green cluster. At this point, even if the second selected pivot edge is a green one, say (X,Z), ChromaticQwickCluster would form only a small green cluster {X,W,Z}.

Motivated by the previous example, Bonchi et al. [2012] introduce the LazyChromaticQwickCluster heuristic, which tries to minimize the risk of bad choices. Given a vertex $x \in V$, and a label $l \in L$, let $\delta(x, l)$ denote the number of edges incident on x having label l. Also, let $\Delta(x) = \max_{l \in L} \delta(x, l)$, and $\lambda(x) = \mathrm{argmax}_{l \in L} \delta(x, l)$. LazyChromaticQwickCluster differs from ChromaticQwickCluster in two aspects:

Pivot random selection: At each iteration LazyChromaticQwickCluster selects a pivot edge in two steps. First, a vertex x is picked up with probability directly proportional to $\Delta(x)$. Then, a second vertex y is selected among the neighbors of x with probability proportional to $\delta(y, \lambda(x))$.

Cluster formation: Given the pivot (x, y), ChromaticQwickCluster forms a cluster by adding all vertices z such that $\langle x, y, z \rangle$ is a monochromatic triangle. LazyChromaticQwickCluster instead, iteratively adds vertices z in the cluster as long as they form a triangle $\langle X, Y, z \rangle$ of color $\ell(x, y)$, where X is either x or y, and Y can be any other vertex already belonging to the current cluster.

Note that the heuristics at the basis of LazyChromaticQwickCluster require no input parameters: the probabilities of picking a pivot are indeed defined based on $\Delta(\cdot)$ and $\delta(\cdot, \cdot)$, which are properties of the input graph. The algorithm is therefore parameter-free like ChromaticQwickCluster. The details of LazyChromaticQwickCluster are reported as Algorithm 4.17, while in the following example we further explain how the LazyChromaticQwickCluster actually works.

Example 4.10   Consider again Figure 4.3. Vertices X and Y have the maximum number of edges of one color: they both have five green edges. Hence, one of them is chosen as the first pivot vertex x by LazyChromaticQwickCluster with higher probability than the remaining vertices. Suppose that X is picked up, i.e., $x = $ X. Given this choice, the second pivot y is chosen among the neighbors of X with probability proportional to $\delta(y, \lambda(x))$, i.e., the higher the number of green edges of the neighbor, the higher the probability for it to be chosen. In this case, hence, LazyChromaticQwickCluster would likely choose Y as a second pivot vertex y, thus making (X,Y) the selected pivot edge. Afterward, LazyChromaticQwickCluster adds to the newly formed cluster the vertices {U,V,Z} because each of them forms a green triangle with the pivot edge. Then, R enters the cluster too, because it forms a green triangle with Y and V, which is already in the cluster. Similarly, W enters the cluster thanks to Z.

The time complexity of the LazyChromaticQwickCluster algorithm is $\mathcal{O}((|L| + \log |V|) |E|)$ (opposed to the $\mathcal{O}(|V| + |E|)$ time complexity of ChromaticQwickCluster).

Anava et al. [2015] also develop a fast heuristic algorithm that is motivated by scenarios in which the input instance admits a ground-truth clustering. Indeed, in many practical settings

---

Algorithm 4.17 LazyChromaticQwickCluster

Input: Edge-labeled graph $G = (V, E, L, l_0, \ell)$, where $\ell : V_2 \to L \cup \{l_0\}$;
Output: Clustering $\mathcal{C} : V \to \mathbb{N}$; cluster labeling function $\ell : \mathcal{C}[V] \to L$

1: $i \leftarrow 1$
2: **while** $E \neq \emptyset$ **do**
3:      pick a random vertex $x \in V$ with probability proportional to $\Delta(x)$
4:      pick a random vertex $y \in \{z \in V \mid (x, z) \in E\}$ with probability proportional to $\delta(y, \lambda(x))$

5:      $C \leftarrow \{x, y\} \cup \{z \in V \mid \ell(x, z) = \ell(y, z) = \ell(x, y)\}$
6:      **repeat**
7:          $C' \leftarrow \{z \in V \mid \exists w \in C \land \ell(x, z) = \ell(w, z) = \ell(x, w)\}$
8:          $C'' \leftarrow \{z \in V \mid \exists w \in C \land \ell(y, z) = \ell(w, z) = \ell(y, w)\}$
9:          $C \leftarrow C \cup C' \cup C''$
10:     **until** $C' \cup C'' = \emptyset$
11:     $\mathcal{C}(x) \leftarrow i$, for all $x \in C$
12:     $\ell(i) = \ell(u, v)$
13:     remove C from G: $V \leftarrow V \setminus C$, $E \leftarrow E \setminus \{(x, y) \in E \mid x \in C\}$
14:     $i \leftarrow i + 1$
15: **end while**
16: **for all** $x \in V$ **do**
17:     $\mathcal{C}(x) \leftarrow i$
18:     $\ell(i) \leftarrow$ a label from L
19:     $i \leftarrow i + 1$
20: **end for**

---

of chromatic correlation clustering, there is an unknown but correct way to partition the objects into clusters. Note that although this underling true clustering exists, it may not be readily identifiable from the data due to noisy similarity observations (similar assumptions have been suggested and utilized by other authors, e.g., Ailon and Liberty [2009], Joachims and Hopcroft [2005], Mathieu and Schudy [2010]; see Sections 1.7 and 4.6).

The proposed heuristic follows the same intuition and a similar outline of the approximated framework presented in Section 4.2.1. Initially, the algorithm modifies the input graph by negating all the positive edges incident on each vertex v whose color is different from the leading color of v. It continues by picking a pivot vertex uniformly at random, and creating a cluster that consists of this pivot and all the adjacent vertices connected to it using a positive edge (i.e., first-level vertices). The algorithm then considers each of the vertices that are connected to the first-level vertices by a positive edge (i.e., second-level vertices), and greedily adds it to the underlying cluster if it improves the post-assignment cost. Once this step is completed, all the clustered vertices are removed from the graph, and the algorithm begins a new iteration by randomly picking a new pivot vertex.

The intuition behind the algorithm is intuitive. Under the assumption that a chromatic-correlation-clustering scenario admits a ground-truth clustering, a major difficulty in identifying this clustering is that the noisy observations obscure the underlying clique structure. The proposed algorithm targets this situation by adding an extra validation step to the basic pivot-based approach which mitigates the noise effects. Recall that the basic approach creates a cluster in each iteration that consists of all the vertices similar to some pivot vertex. The algorithm proposed by Anava et al. [2015] augments this approach by finding additional vertices that, although appearing dissimilar to the pivot due to noise, have high similarity to the other cluster vertices. This heuristic algorithm has time complexity linear in the input size, thus very efficient.

A nice feature of the previous algorithms is that they are parameter-free: they produce clusterings by using information that is local to the pivot edges, without forcing the number of output clusters in any way. However, in some cases, it would be desirable to have a pre-specified number K of clusters. To this end, Bonchi et al. [2012] present an algorithm based on the *alternating-minimization* paradigm [Csiszar and Tusnady, 1984] that takes as input the number K of output clusters and attempts to minimize Equation (4.2) directly. The algorithm starts with a random assignment of both vertices and labels to clusters and works by iteratively alternating between two optimization steps until convergence, i.e., until no changes, neither in C nor in $c\ell$, have been observed with respect to the previous iteration. Both the steps of the algorithm are solved optimally. As a consequence, the value of the CHROMATIC-CORRELATION-CLUSTERING objective function is guaranteed to be non-increasing in every iteration of the algorithm, until convergence. Finding the global optimum is obviously hard, but the algorithm is guaranteed to converge to a local minimum.

Finally, Klodt et al. [2021] devise a greedy heuristic that iteratively chooses an unclustered edge and builds a cluster around it as follows. First, it considers the neighborhood of the edge and temporarily add a sample of it to the current cluster. Then, vertices that do not fit are iteratively removed to produce a good starting cluster. As taking the whole neighborhood can be time consuming for vertices with a high degree, the sample size of neighbored vertices is restricted with a parameter. To increase the chances of finding a good initial cluster, vertices from the common neighborhood of the starting edge of the sample are preferred. Then, the initial cluster is greedily expanded with the vertices that maximize the merge gain.

4.3   MULTILAYER GRAPHS

In social media and social networks, as well as in several other real-world contexts—such as biological and financial networks, transportation systems, and critical infrastructures—there might be multiple types of relation among entities. Data in these domains is typically modeled as a *multilayer network* (also known as multidimensional network), i.e., a graph where multiple edges of different types may exist between any pair of vertices [Dickison et al., 2016, Galimberti et al., 2017, 2020, Lee et al., 2015, Tagarelli et al., 2017].

In this context, Bonchi et al. [2015] generalize the framework introduced in the previous section (Section 4.2) to deal with edge-labeled graphs where each edge can have more than one label. This generalization of Problem 4.3 is achieved by (i) measuring the intra-cluster label homogeneity by means of a distance function d_ℓ between sets of labels, and (ii) allowing the output $c\ell$ cluster-labeling function to assign *a set* of labels to each cluster (instead of a single label). For the sake of presentation, in Problem 4.11 and in the remainder of this section, we assume that the powerset 2^L does not include the empty set, i.e., $2^L = \{S \subseteq L \mid |S| \geq 1\}$.

Problem   4.11   (Multi-Chromatic-Correlation-Clustering) Let $G = (V, E, L, l_0, \ell)$ be a multi-layer graph, where V is a set of vertices, $E \subseteq V_2$ is a set of edges, L is a set of labels, $l_0 \notin L$ is a *special* label, and $\ell : V_2 \to 2^L \cup \{l_0\}$ is a labeling function that assigns a set of labels to each unordered pair of vertices in V such that $\ell(x, y) = l_0$ if and only if $(x, y) \notin E$. Let also $d_\ell : 2^L \cup \{l_0\} \times 2^L \cup \{l_0\} \to \mathbb{R}^+$ be a distance function between sets of labels. Find a clustering $\mathcal{C} : V \to \mathbb{N}$ and a cluster-labeling function $\ell : \mathcal{C}[V] \to 2^L$ so as to minimize the cost

$$\text{cost}(G, \mathcal{C}, \ell) = \sum_{\substack{(x,y) \in V_2, \\ \mathcal{C}(x) = \mathcal{C}(y)}} d_\ell(\ell(x, y), \ell(\mathcal{C}(x))) + \sum_{\substack{(x,y) \in V_2, \\ \mathcal{C}(x) \neq \mathcal{C}(y)}} d_\ell(\ell(x, y), \{l_0\}). \quad (4.4)$$

It is easy to see that MULTI-CHROMATIC-CORRELATION-CLUSTERING is a generalization of CHROMATIC-CORRELATION-CLUSTERING. In particular, MULTI-CHROMATIC-CORRELATION-CLUSTERING reduces to CHROMATIC-CORRELATION-CLUSTERING when all edges in the input graph are single-labeled and the distance d_ℓ is defined as

$$d_\ell(\{l_1\}, \{l_2\}) = \begin{cases} 0, & \text{if } l_1 = l_2 \\ 1, & \text{otherwise.} \end{cases}$$

A key point of differentiation between the multiple-label formulation in Bonchi et al. [2015] and single-label formulation in Bonchi et al. [2012] is that the former employs a distance function d_ℓ to measure how much label sets differ from each other. Among the various possible choices of d_ℓ, Bonchi et al. [2015] adopt the popular Hamming distance, as a good tradeoff between simplicity and effectiveness. The Hamming distance between two sets of labels $L_1 \subseteq L$, $L_2 \subseteq L$ is defined as the number of disagreements between L_1 and L_2:

$$d_\ell(L_1, L_2) = |L_1 \setminus L_2| + |L_2 \setminus L_1|.$$

In this context, thus, the distance d_ℓ lies within the range $[0 \ldots |L| + 1]$. Also, this choice of distance makes the cost paid by an inter-cluster edge in Equation (4.4) equal to the number of labels present on that edge (plus 1), i.e., $d_\ell(\ell(x, y), \{l_0\}) = |\ell(x, y)| + 1$, $\forall (x, y) \in E$. This is desirable, as we recall that in the ideal output clustering every pair of vertices in different clusters should have a relation represented by the l_0 label, which models the "no-edge" case. Thus, the cost of an inter-cluster edge should be a function of the size of the label set present on that edge: the larger the number of labels, the further that relation from the absent-edge case.

4.3.1   ALGORITHMS

Bonchi et al. [2015] present an approximation algorithm for solving the MULTI-CHROMATIC-CORRELATION-CLUSTERING problem (Problem 4.11). The proposed algorithm, called Multi-ChromaticQwickCluster, roughly follows the scheme of the ChromaticQwickCluster algorithm designed for the single-label version of the problem.

Algorithm 4.18 Multi-ChromaticQwickCluster

Input: Edge-labeled graph $G = (V, E, L, l_0, \ell)$, where $\ell : V_2 \to 2^L \cup \{l_0\}$
Output: Clustering $\mathcal{C} : V \to \mathbb{N}$; cluster labeling function $\ell : \mathcal{C}[V] \to 2^L$

   $i \leftarrow 1$
   while $E \neq \emptyset$ **do**
      pick an edge $(x, y) \in E$ uniformly at random
      $C \leftarrow \{x, y\} \cup \{z \in V \mid d_\ell(\ell(x, y), \ell(x, z)) = d_\ell(\ell(x, y), \ell(y, z)) = 0\}$
      $\mathcal{C}(x) \leftarrow i$, for all $x \in C$
      $\ell(i) = \ell(x, y)$
      $V \leftarrow V \setminus C$, $E \leftarrow E \setminus \{(x, y) \in E \mid x \in C\}$ (remove C from G)
      $i \leftarrow i + 1$
   end while
   for all $x \in V$ **do**
      $\mathcal{C}(x) \leftarrow i$
      $\ell(i) \leftarrow$ a label set from 2^L
      $i \leftarrow i + 1$
   end for

The outline of the Multi-ChromaticQwickCluster algorithm is reported as Algorithm 4.18. Like ChromaticQwickCluster, the main idea of the algorithm is to pick a pivot edge (x, y) uniformly at random, build a cluster around it, and remove such a cluster from the graph. The process continues until there is no edge remaining to be picked as a pivot. A major peculiarity of Multi-ChromaticQwickCluster is the presence of multiple labels on the edges of the input graph, which implies that the distance function d_ℓ between sets of labels should now be taken into account to determine how the cluster is built around the pivot (x, y). In particular, the cluster C defined at every iteration of the algorithm is composed of all vertices z such that the labels on the edges of the triangle (x, y, z) have all distance d_ℓ equal to zero, i.e., $d_\ell(\ell(x, y), \ell(x, z)) = d_\ell(\ell(x, y), \ell(y, z)) = 0$. According to the choice of distance d_ℓ (i.e., Hamming distance), the latter condition is equivalent to having label sets $\ell(x, y), \ell(x, z), (y, z)$ equal to each other, i.e., $\ell(x, y) = \ell(x, z) = \ell(y, z)$. Moreover, the presence of multiple labels on edges affects the way how the output clusters are labeled: following the common intuition, all clusters can now have more than one label assigned. Finally, the time complexity of Multi-ChromaticQwickCluster is $\mathcal{O}(|L| \times |E|)$.

The Multi-ChromaticQwickCluster achieves a provable approximation guarantee, as formally stated in the following theorem.

Theorem 4.12 [Bonchi et al., 2015]. *The approximation ratio of the* Multi-ChromaticQwickCluster *algorithm on input* $G = (V, E, L, l_0, \ell)$ *is*

$$r(G) = \frac{\text{cost}(G)}{\text{cost}^*(G)} \leq 3|L|T_{max}.$$

Proof. The MULTI-CHROMATIC-CORRELATION-CLUSTERING problem formulation can be reinterpreted from a single-label perspective by considering all labels in 2^L as a "single" label. The difference from the standard CHROMATIC-CORRELATION-CLUSTERING formulation is that the cost paid by a single vertex pair is this way bounded by $|L|$ (rather than being ≤ 1). Based on this observation, it is easy to see that running the Multi-ChromaticQwickCluster algorithm on an instance G of the MULTI-CHROMATIC-CORRELATION-CLUSTERING problem is equivalent to running the ChromaticQwickCluster algorithm on the same instance G reinterpreted from a single-label perspective. This means that a vertex pair (x, y) in the Multi-ChromaticQwickCluster solution pays a non–zero cost only if the cost paid by the same vertex pair in the solution output by ChromaticQwickCluster on the single-label reinterpretation of the problem instance is greater than zero. At the same time, however, the cost of (x, y) in the Multi-ChromaticQwickCluster solution is bounded by $|L|$. Combining the latter two arguments implies that the approximation ratio of Multi-ChromaticQwickCluster is guaranteed to be within a factor $|L|$ of the approximation ratio achieved by the ChromaticQwickCluster algorithm: the approximation ratio of Multi-ChromaticQwickCluster is thus $3|L|T_{max}$.   □

The above theorem states that the approximation ratio of Multi-ChromaticQwickCluster increases the approximation ratio of ChromaticQwickCluster by a factor $|L|$. However, this is not really problematic as the number of labels on real-world edge-labeled graphs is typically small and can safely be assumed to be constant. Moreover, the approximation ratio remains constant for bounded-degree graphs.

4.4 VERTEX-LABELED GRAPHS

Ahmadian et al. [2020] study the problem of *Fair Correlation Clustering* defined over a vertex-labeled graph, where each vertex is assigned one and only one label (color). The idea is that the vertex label represents some protected attribute, such as gender or ethnicity, and the clustering is constrained to provide a fair representation of each class of vertex in every cluster. In particular, Ahmadian et al. [2020] consider a fairness constraint requiring that each color constitutes at most an α-fraction of each clustering for a given threshold $\alpha \in (0, 1)$, and study three specific cases: $\alpha = 1/2$, $\alpha = 1/|L|$ where L is the set of possible vertex colors, and more generally $\alpha = 1/t$ where t is any positive integer. We next define the problem under the latter, more general, constraint.

> **Problem 4.13 (Fair-Correlation-Clustering)** Let $G = (V, E, L, \ell)$ be a vertex-labeled graph, where V is a set of vertices, $E \subseteq V_2$ is a set of edges, L is a set of vertex labels, $\ell : V \to L$ is a labeling function that assigns one label to each vertex. Given a threshold $\alpha = 1/t$ where $t \in \mathbb{Z}^+$, find a clustering $\mathcal{C} : V \to \mathbb{N}$ such that each color constitutes at most an α-fraction of each clustering and, among all the clusterings satisfying this fairness constraint, it minimizes the (standard) correlation-clustering min-disagreement objective.

A max-agreement version can also be defined similarly. However, Ahmadian et al. [2020] focus on the minimization counterpart only, as the maximization version admits a trivial randomized 2-approximation that can be made fair.

Ahmadian et al. [2020] build on previous work by Chierichetti et al. [2017], who provide results for fair versions of k-center and k-median clustering problems. Chierichetti et al. [2017] develop their work upon the notion *fairlet*, i.e., a small set of elements that satisfies the fairness property, Specifically, they show that fair k-median and k-center can be solved by first decomposing an instance into fairlets and then solving the clustering problem on the set of centers of those fairlets.

Ahmadian et al. [2020] devise a fairlet-based reduction for correlation clustering. Although in the case of k-center and k-median the fairlet-decomposition problem amounts to solving the same clustering problem on the same instance under the condition that each cluster is a fairlet, the situation for correlation clustering is complicated by the lack of properties of metric spaces. To tackle this challenge, Ahmadian et al. [2020] introduce a novel cost function for the correlation-clustering fairlet decomposition, and prove that this cost can be approximated by a median-type clustering cost function for a carefully defined metric space. Given a solution to this fairlet-decomposition problem, Ahmadian et al. [2020] show that the FAIR-CORRELATION-CLUSTERING instance can be reduced to a regular correlation-clustering instance through a graph transformation. Therefore, any approximation algorithm for fairlet decomposition with median cost yields an approximation algorithm for FAIR-CORRELATION-CLUSTERING; the loss in the approximation ratio depends on the size of the fairlets. However, Ahmadian et al. [2020] also show that, in many natural cases, there is a fairlet decomposition with a small number of fairlets, thereby bounding the approximation ratio of their algorithm. In particular, Ahmadian et al. [2020] define approximation algorithms for the cases $\alpha = 1/2$ and $\alpha = 1/|L|$, and a bicriteria approximation algorithm for the case $\alpha = 1/t$, with upper bounds of 3, $|L|$, and $2t - 1$, respectively, on the size of fairlets. We next state Ahmadian et al. [2020]'s ultimate result for the most general setting.

Theorem 4.14 [Ahmadian et al., 2020]. *For $\alpha = 1/t$, given a γ-approximation algorithm for fairlet decomposition with median cost, there is an $\mathcal{O}(t\gamma)$-approximation algorithm for fair correlation clustering.*

For the details of the algorithms and their analysis, we refer the interested reader to the original paper by Ahmadian et al. [2020].

4.5 HYPERGRAPHS

Li et al. [2017] (who we already mentioned in Section 3.1 about OVERLAPPING-CORRELATION-CLUSTERING) introduce a higher-order generalization of correlation clustering, which they call MIXED-MOTIF-CORRELATION-CLUSTERING. This problem is introduced as a means for clustering networks based on higher-order motif patterns shared among vertices, and is motivated by previous successful results on motif-based graph clustering (see, e.g., Benson et al. [2016]). Although a similar higher-order correlation-clustering problem was considered by Kim et al. [2011] for image segmentation, Li et al. [2017] were the first to study it from a theoretical perspective.

The MIXED-MOTIF-CORRELATION-CLUSTERING problem is as follows. Let E_k denote the set of all k-tuples of vertices (hyperedges of size k) in G, and let each $\mathcal{E} \in E_k$ have a positive weight, $w_{\mathcal{E}}^+$, and a negative weight, $w_{\mathcal{E}}^-$. If a clustering separates at least one pair of vertices in $\mathcal{E}$, this gives a penalty of $w_{\mathcal{E}}^+$; otherwise, if all vertices in $\mathcal{E}$ are clustered together, there is a penalty of $w_{\mathcal{E}}^-$. Formally:

Problem 4.15 (Mixed-Motif-Correlation-Clustering) Let $G = (V, E)$ be a hypergraph, where E is the union of all sets E_k, with $2 \leq k \leq |V|$, E_k denotes the set of all k-tuples of vertices (hyperedges of size k) in G, and let each $\mathcal{E} \in E_k$ have a positive weight, $w_{\mathcal{E}}^+$, and a negative weight, $w_{\mathcal{E}}^-$. Find a clustering $\mathcal{C} : V \to \mathbb{N}$ that minimizes

$$\sum_{k=2}^{|V|} \lambda_k \sum_{\substack{\mathcal{E} \in E_k, \\ \mathcal{E} \subseteq C_i \text{ for some cluster } C_i \text{ of } \mathcal{C}}} w_{\mathcal{E}}^+ + \lambda_k \sum_{\substack{\mathcal{E} \in E_k, \\ \mathcal{E} \nsubseteq C_i \text{ for some cluster } C_i \text{ of } \mathcal{C}}} w_{\mathcal{E}}^-,$$

where $\lambda_k \geq 0$ defines the relevance factor for hyperedges of size k.

Setting $\lambda_2 = 1$ (edges) and $\lambda_k = 0, \forall k > 2$, we obtain the standard min-disagreement formulation of correlation clustering. Li et al. [2017] show that the problem is **NP**-completeand provide an LP formulation. They limit their theoretical analysis to the case of hyperedges of sizes 2 and 3 only, when hyperedge weights satisfy probability constraints ($w_{\mathcal{E}}^+ + w_{\mathcal{E}}^- = 1$, for every hyperedge $\mathcal{E}$ of size 2 or 3), and assuming $\lambda_2 = \lambda_3 = 1$. For this setting, they show a 9-approximation.

By rounding the same LP formulation as the one introduced by Li et al. [2017], Fukunaga [2018] provides an $\mathcal{O}(k \log n)$ approximation for general weighted hypergraphs.

Gleich et al. [2018] generalize the 4-approximation of Charikar et al. [2005] for complete unweighted graphs to obtain a $4(k-1)$-approximation for MIXED-MOTIF-CORRELATION-CLUSTERING on hypergraphs with hyperedges of fixed dimension k and satisfying probability constraints. They consider the same LP relaxation as Li et al. [2017], and apply a similar rounding technique. However, they provide an approximation guarantee for arbitrary k that is linear in k, while additionally improving the factor for $k = 3$ from 9 to 8.

Gleich et al. [2018] also study a variant of MIXED-MOTIF-CORRELATION-CLUSTERING where only two clusters are sought, termed 2-MIXED-MOTIF-CORRELATION-CLUSTERING. They design an algorithm that gives an asymptotic $(1 + k\,2^{k-2})$-approximation, by generalizing the 3-approximation of Bansal et al. [2004] for MIN-DISAGREE[2] (see Section 2.1). This is the first combinatorial result for 2-MIXED-MOTIF-CORRELATION-CLUSTERING, and is a 7-approximation for $k = 3$.

One last contribution by Gleich et al. [2018] is a discussion of several interesting open questions. First, whether an approximation that is independent of k could be developed for minimizing disagreement in hypergraphs. Another interesting question is whether a pivoting algorithm à la Ailon et al. [2008a] could be developed for the MIXED-MOTIF-CORRELATION-CLUSTERING problem. As far as agreement maximization, the simple strategy of either placing all the vertices in a single cluster or making all the vertices singletons still leads to a $\frac{1}{2}$-approximation for hypergraphs with arbitrary weights and any k. This leads to open questions about what results for agreement maximization can be generalized to the hypergraph setting.

4.6   NOISY GRAPHS

One natural way of thinking about correlation-clustering problem instances is to assume that a ground-truth clustering exists, but the observations that we have of it (our input) are noisy. In this regard, it is of interest to ask how well an algorithm can reconstruct the ground truth for different levels of noise in the input.

Already in the seminal paper by Bansal et al. [2004], a discussion on the behavior of their algorithm under noise in the data was presented. However, the first formal theoretical analysis of this task is due to Joachims and Hopcroft [2005]. They give bounds on the error with which correlation clustering recovers the ground truth under a simple probabilistic model over graphs that extends the well-known *planted-partition model*. Furthermore, Joachims and Hopcroft [2005] study the asymptotic behavior with respect to the density of the graph and the scaling of cluster sizes. Finally, they propose a statistical test for evaluating the significance of a clustering.

Mathieu and Schudy [2010] study a semi-random model for correlation clustering on complete graphs with unit edge costs. The model is as follows. Start from an arbitrary partition $\mathcal{B}$ of the n input vertices into clusters (base clustering). Then, each pair of vertices is perturbed independently with probability p. In the fully-random model, the input is generated from $\mathcal{B}$ simply by switching every perturbed pair. In the semi-random variant, an adversary controls the perturbed pairs and decides whether to switch them or not. Mathieu and Schudy [2010] propose an algorithm based on semidefinite programming (SDP relaxations with ℓ_2^2-triangle inequality constraints) for their semi-random model on complete graphs. It finds a clustering of cost at most $1 + \mathcal{O}(n^{-1/6})$ times the cost of the optimal clustering (as long as $p \leq 1/2 - \mathcal{O}(n^{-1/3})$) and manages to approximately recover the ground-truth solution (when the clusters have at least a certain size, specifically in the order of $\sqrt{n}$).

Chen et al. [2014] extend the framework from complete graphs to sparser Erdös–Rényi random graphs. In their model, the underlying unlabeled graph $G(V, E)$ comes from an Erdös–Rényi random graph (of edge probability p), the label of each edge is set (independently) to be consistent with the ground-truth clustering with probability $1 - \varepsilon$ and inconsistent with probability ε. Using weaker convex relaxations, Chen et al. [2014] obtain an algorithm that recovers the ground truth when $p \geq k^2 \log^{O(1)} n / n$.

While the average-case models of Mathieu and Schudy [2010] and Chen et al. [2014] are natural, they are unrealistic in practice since most real-world graphs are neither dense nor captured by Erdös–Rényi distributions. Further, these models assume that every pair of vertices have the same amount of similarity or dissimilarity (all costs are unitary). Makarychev et al. [2015] overcome these limitations by introducing a semi-random model which assumes very little about the observations, while also allowing non-uniform costs. The proposed model generalizes the model of Mathieu and Schudy [2010] by considering a semi-random instance $\{G(V, E, c), (E_+, E_-)\}$ to be generated as follows.

1. The adversary chooses an undirected graph $G(V, E, c)$ and a partition $\mathcal{P}^*$ of the vertex set V (referred to as the planted clustering or ground-truth clustering).

2. Every edge in E is included in set E_R independently with probability ε.

3. Every edge $(u, v) \in E \setminus E_R$ with u and v in the same cluster of $\mathcal{P}^*$ is included in E_+, and every edge $(u, v) \in E \setminus E_R$, with u and v in different clusters of $\mathcal{P}^*$ is included in E_-.

4. The adversary adds every edge from E_R either to E_+ or to E_- (but not to both sets).

Makarychev et al. [2015] give two approximation algorithms for correlation clustering instances from this model. The first algorithm finds a solution of cost $(1 + \delta)\mathsf{OPT} + \mathcal{O}_\delta(n \log^3 n)$ with high probability (for every $\delta > 0$), where OPT is the value of the optimal solution to the input problem instance. The second algorithm finds the ground-truth clustering with an arbitrarily small classification error η (under some additional assumptions on the input problem instance).

CHAPTER 5

Other Computational Settings

This chapter overviews the work on Correlation Clustering in special contexts, which carry over additional challenges with respect to more conventional settings.

5.1 QUERY-EFFICIENT CORRELATION CLUSTERING

Despite its considerable appeal, one practical drawback of correlation clustering is the fact that, given n items to be clustered, $\Theta(n^2)$ similarity computations are needed to prepare the similarity graph that serves as input for the algorithms. In addition to the obvious algorithmic cost involved with $\Theta(n^2)$ queries, in certain applications there is an additional type of cost that may render correlation clustering algorithms impractical. Consider the following motivating real-world scenarios. In biological sciences, in order to produce a network of interactions between a set of biological entities (e.g., proteins) a highly trained professional has to devote time and costly resources (e.g., equipment) to perform tests between all $\binom{n}{2}$ pairs of entities. In entity resolution, a task central to data integration and data cleaning [Wang et al., 2012], a crowdsourcing-based approach performs queries to workers of the form "does the record x represent the same entity as the record y?" Such queries to workers involve a monetary cost, hence it is desirable to reduce it. In both scenarios, developing clustering tools that use less than $\binom{n}{2}$ queries is of major interest. At a high level, we wish to answer the following question.

> **Problem 5.1** Design a correlation clustering algorithm that outputs a good approximation in a *query-efficient* manner: given a budget Q of queries, the algorithm is allowed to learn the specific value of $s(i, j) \in \{0, 1\}$ for each query (i, j) of the algorithm's choice.

The study of query-efficient algorithms for correlation clustering is relatively new. Ailon et al. [2012b] obtain algorithms for minimizing disagreements with sublinear query complexity, but their running time is exponential in n. The first query-efficient algorithm which is also computationally efficient was designed by Bonchi et al. [2013a] and García-Soriano et al. [2020]. We present their main findings here.

The Query-Efficient-Correlation-Clustering (for short, QECC) algorithm by García-Soriano et al. [2020] focuses on the MIN-DISAGREE variant of correlation clustering, and consists in simply running the well-established QwickCluster algorithm [Ailon et al., 2008a] (see Section 1.5) until the query budget Q has been reached, and outputting singleton clusters for the remaining unclustered vertices (Algorithm 5.19).

Algorithm 5.19 QECC

Input: Unweighted graph $G = (V, E)$; query budget Q
Output: Clustering of G
   $R \leftarrow V$ // Unclustered vertices so far
   while $|R| > 0 \wedge Q \geq |R| - 1$ **do**
      Pick a pivot v from R uniformly at random.
      Query all pairs (v, w) for $w \in R \setminus v$ to determine $\Gamma_G^+(v) \cap R$.
      $Q \leftarrow Q - |R| + 1$
      Output cluster $C = \{v\} \cup \Gamma_G^+(v) \cap R$.
      $R \leftarrow R \setminus C$
   end while
   Output a separate singleton cluster for each remaining $v \in R$.

5.1.1   ANALYSIS OF QECC

Upper Bound

The main result of García-Soriano et al. [2020] about the QECC algorithm is as follows.

Theorem 5.2   [García-Soriano et al., 2020].   *Let G be a graph with n vertices. For any $Q > 0$, the QECC algorithm (Algorithm 5.19) finds a clustering of G with expected cost at most $3\mathsf{OPT} + \frac{n^3}{2Q}$ (where OPT is the cost of the optimal solution), making at most Q edge queries, and running in $\mathcal{O}(Q)$ time (under the assumption of unit-cost queries).*

In the remainder of this subsection we show how García-Soriano et al. [2020] got this result. For the sake of simplicity, we hereinafter identify a complete "+,-" labeled graph G with its graph of *positive* edges (V, E^+), so that queries correspond to querying a pair of vertices for the existence of an edge. The set of (positive) neighbors of v in a graph $G = (V, E)$ will be denoted $\Gamma(v)$; a similar notation is used for the set $\Gamma(S)$ of positive neighbors of a set $S \subseteq V$. The cost of the optimum clustering for G is denoted OPT. When ℓ is a clustering, $\mathrm{cost}(\ell)$ denotes the cost (number of disagreements) of this clustering, defined by (1.1) with $s(x, y) = 1$ iff $\{x, y\} \in E$.

In order to analyze QECC, we need to understand how early stopping of QwickCluster affects the accuracy of the clustering found. For any non-empty graph G and pivot $v \in V(G)$, let $N_v(G)$ denote the subgraph of G resulting from removing all edges incident to $\Gamma(v)$ (keeping all vertices). Define a random sequence $G_0, G_1, \ldots$ of graphs by $G_0 = G$ and $G_{i+1} = N_{v_{i+1}}(G_i)$, where $v_1, v_2, \ldots$ are chosen independently and uniformly at random from $V(G_0)$. Note that $G_{i+1} = G_i$ if at step i a vertex is chosen for a second time.

The following lemma is key.

Lemma 5.3 [García-Soriano et al., 2020] *Let G_i have average degree $\tilde{d}$. When going from G_i to G_{i+1}, the number of edges decreases in expectation by at least $\binom{\tilde{d}+1}{2}$.*

Proof. Let $V = V(G_0)$, $E = E(G_i)$ and let $d_u = |\Gamma(u)|$ denote the degree of $u \in V$ in G_i. Consider an edge $\{u, v\} \in E$. It is deleted if the chosen pivot v_i is an element of $\Gamma(u) \cup \Gamma(v)$ (which contains u and v); let X_{uv} be the 0-1 random variable associated with this event. It occurs with probability

$$\mathbb{E}[X_{uv}] = \frac{|\Gamma(u) \cup \Gamma(v)|}{n} \geq \frac{1 + \max(d_u, d_v)}{n} \geq \frac{1}{n} + \frac{d_u + d_v}{2n}.$$

Let $D = \sum_{u<v | \{u,v\} \in E} X_{uv}$ be the number of edges deleted (we assume an ordering of V to avoid double-counting edges). By linearity of expectation,

$$
\begin{aligned}
\mathbb{E}[D] &= \sum_{\substack{u<v \\ \{u,v\} \in E}} \mathbb{E}[X_{uv}] = \frac{1}{2} \sum_{\substack{u,v \in V \\ \{u,v\} \in E}} \mathbb{E}[X_{uv}] \\
&\geq \frac{1}{2} \sum_{\substack{u,v \\ \{u,v\} \in E}} \left(\frac{1}{n} + \frac{d_u + d_v}{2n} \right) \\
&= \frac{\tilde{d}}{2} + \frac{1}{4n} \sum_{\substack{u,v \\ \{u,v\} \in E}} (d_u + d_v).
\end{aligned}
$$

Now we compute

$$
\begin{aligned}
\frac{1}{4n} \sum_{\substack{u,v \\ \{u,v\} \in E}} (d_u + d_v) &= \frac{1}{2n} \sum_{\substack{u,v \\ \{u,v\} \in E}} d_u = \frac{1}{2n} \sum_u d_u^2 \\
&= \frac{1}{2} \mathbb{E}_{u \sim V}[d_u^2] \geq \frac{1}{2} \left(\mathbb{E}_{u \sim V}[d_u] \right)^2 = \frac{1}{2} \tilde{d}^2,
\end{aligned}
$$

where we used the Cauchy–Schwarz inequality in the last line. Hence, $\mathbb{E}[D] \geq \frac{\tilde{d}}{2} + \frac{\tilde{d}^2}{2} = \binom{\tilde{d}+1}{2}$. $\square$

Lemma 5.4 [García-Soriano et al., 2020] *Let G be a graph with n vertices and let $P = \{v_1, \ldots, v_r\}$ be the first r pivots chosen by running QwickCluster on G. Then the expected number of positive edges of G not incident with an element of $P \cup \Gamma(P)$ is less than $\frac{n^2}{2(r+1)}$.*

Proof. Recall that at each iteration QwickCluster picks a random pivot from R. This selection is equivalent to picking a random pivot v from the original set of vertices V and discarding it if $v \notin R$, repeating until some $v \in R$ is found, in which case a new pivot is added. Consider the following modification of QwickCluster, denoted SluggishCluster, which picks a pivot v at random from V but always increases the counter r of pivots found, even if $v \in R$ (ignoring the cluster creation step if $v \notin R$). We can couple both algorithms into a common probability space where each point ω contains a sequence of randomly selected vertices and each algorithm picks the next one in sequence. For any ω, whenever the first r pivots of SluggishCluster are $S = (v_1, \ldots, v_r)$, then the first r' pivots of QwickCluster are the sequence S' obtained from S by removing previously appearing elements, where $r' = |S'|$. Hence, $|V \setminus (S \cup \Gamma(S))| = |V \setminus (S' \cup \Gamma(S'))|$ and $r' \leq r$. Thus, the number of edges not incident with the first r pivots and their neighbors in SluggishCluster stochastically dominates the number of edges not incident with the first r pivots and their neighbors in SluggishCluster, since both numbers are decreasing with r.

Therefore, it is enough to prove the claim for SluggishCluster. Let $n = |V(G_0)|$ and define $\alpha_i \in [0, 1]$ by $\alpha_i = \frac{2 \cdot |E(G_i)|}{n^2}$. We claim that for all $i \geq 1$ the following inequalities hold:

$$\mathbb{E}[\alpha_i \mid G_0, \ldots, G_{i-1}] \leq \alpha_{i-1}(1 - \alpha_{i-1}), \tag{5.1}$$

$$\mathbb{E}[\alpha_i] \leq \mathbb{E}[\alpha_{i-1}] \left(1 - \mathbb{E}[\alpha_{i-1}]\right), \tag{5.2}$$

$$\mathbb{E}[\alpha_i] < \frac{1}{i+1}. \tag{5.3}$$

Indeed, G_i is a random function of G_{i-1} only, and the average degree of G_{i-1} is $\widetilde{d}_{i-1} = \alpha_{i-1} n$ so, by Lemma 5.3,

$$\mathbb{E}[2 \cdot |E(G_i)| \mid G_{i-1}] \leq \alpha_{i-1} n^2 - 2 \cdot \frac{1}{2} \widetilde{d}_{i-1}^2 = n^2 \alpha_{i-1}(1 - \alpha_{i-1}),$$

proving (5.1). Now (5.2) now follows from Jensen's inequality: since

$$\mathbb{E}[\alpha_i] = \mathbb{E}\left[\mathbb{E}[\alpha_i \mid G_0, \ldots, G_{i-1}]\right] \leq \mathbb{E}[\alpha_{i-1}(1 - \alpha_{i-1})]$$

and the function $g(x) = x(1 - x)$ is concave in $[0, 1]$, we have

$$\mathbb{E}[\alpha_i] \leq \mathbb{E}[g(\alpha_{i-1})] \leq g(\mathbb{E}[\alpha_{i-1}]) = \mathbb{E}[\alpha_{i-1}](1 - \mathbb{E}[\alpha_{i-1}]).$$

Finally, we prove $\mathbb{E}[\alpha_i] < 1/(i+1)$ $\forall i \geq 1$. For $i = 1$, we have:

$$\mathbb{E}[\alpha_1] \leq g(\alpha_0) \leq \max_{x \in [0,1]} g(x) = g\left(\frac{1}{2}\right) = \frac{1}{4} < \frac{1}{2}.$$

For $i > 1$, observe that g is increasing on $[0, 1/2]$ and

$$g\left(\frac{1}{i}\right) = \frac{1}{i} - \frac{1}{i^2} \leq \frac{1}{i} - \frac{1}{i(i+1)} = \frac{1}{i+1},$$

so (5.3) follows from (5.2) by induction on i:

$$\mathbb{E}[\alpha_{i-1}] < \frac{1}{i} \implies \mathbb{E}[\alpha_i] \leq g\left(\frac{1}{i}\right) \leq \frac{1}{i+1}.$$

Therefore, $\mathbb{E}[|E(G_r)|] = \frac{1}{2}\mathbb{E}[\alpha_r]n^2 \leq \frac{n^2}{2(r+1)}$, as we wished to show.   □

We are now ready to prove Theorem 5.2.

Proof of Theorem 5.2.   Let OPT denote the cost of the optimal clustering of G and let C_r be a random variable denoting the clustering obtained by stopping QwickCluster after r pivots are found (or running it to completion if it finds r pivots or less), and putting all unclustered vertices into singleton clusters. Note that whenever C_i makes a mistake on a negative edge, so does C_j for $j \geq i$; on the other hand, every mistake on a positive edge by C_i is either a mistake by C_j ($j \geq i$) or the edge is not incident to any of the vertices clustered in the first i rounds. By Lemma 5.4, there are at most $\frac{n^2}{2(i+1)}$ of the latter in expectation. Hence, $\mathbb{E}[\text{cost}(C_i)] - \mathbb{E}[\text{cost}(C_n)] \leq \frac{n^2}{2(i+1)}$.

Algorithm QECC runs for k rounds, where $k \geq \lfloor \frac{Q}{n-1} \rfloor > \frac{Q}{n} - 1$ because each pivot uses $|R| - 1 \leq n - 1$ queries. Then

$$\mathbb{E}[\text{cost}(C_k)] - \mathbb{E}[\text{cost}(C_n)] < \frac{n^2}{2(k+1)} < \frac{n^3}{2Q}.$$

On the other hand, we have $\mathbb{E}[\text{cost}(C_n)] \leq 3 \cdot \text{OPT}$ because of the expected 3-approximation guarantee of QwickCluster from Ailon et al. [2008a]. Thus, $\mathbb{E}[\text{cost}(C_k)] \leq 3\text{OPT} + \frac{n^3}{2Q}$, proving our approximation guarantee.

Finally, the time spent inside each iteration of the main loop is dominated by the time spent making queries to vertices in R, since this number also bounds the size of the cluster found. Therefore, the running time of QECC is $\mathcal{O}(Q)$.   □

Lower Bound

García-Soriano et al. [2020] also show that the QECC algorithm is essentially optimal: for any given budget of queries, no algorithm (adaptive or not) can find a solution better than that of QECC by more than a constant factor.

Theorem 5.5 [García-Soriano et al., 2020].   *For any $c \geq 1$ and $8n < T \leq \frac{n^2}{2048c^2}$, any algorithm finding a clustering with expected cost at most $c \cdot \text{OPT} + T$ must make at least $\Omega(\frac{n^3}{Tc^2})$ adaptive edge similarity queries.*

Proof. Let $\epsilon = \frac{T}{n^2}$; then $\frac{1}{n} < \epsilon \leq \frac{1}{2048c^2}$. By Yao's minimax principle [Yao, 1977], it suffices to produce a distribution $\mathcal{G}$ over graphs with the following properties:

- the expected cost of the optimal clustering of $G \sim \mathcal{G}$ is $\mathbb{E}[\mathrm{OPT}(G)] \leq \frac{\varepsilon n^2}{c}$; and

- for any *deterministic* algorithm making less than $L/2 = \frac{n}{2048\varepsilon c^2}$ queries, the expected cost (over $\mathcal{G}$) of the clustering produced exceeds $2\varepsilon n^2 \geq c \cdot \mathbb{E}[\mathrm{OPT}(G)] + T$.

Let $\alpha = \frac{1}{4c}$ and $k = \frac{1}{32c\varepsilon}$. We can assume that c, k and $\alpha n/k$ are integral (here we use the fact that $\varepsilon > 1/n$). Let $A = \{1, \ldots, (1-\alpha)n\}$ and $B = \{(1-\alpha)n + 1, \ldots, n\}$.

Consider the following distribution $\mathcal{G}$ of graphs: partition the vertices of A into exactly k equal-sized clusters $C_1, \ldots, C_k$. The set of positive edges will be the union of the cliques defined by $C_1, \ldots, C_k$, plus edges joining each vertex $v \in B$ to all the elements of C_{r_v} for a randomly chosen $r_v \in [k]$; r_v is chosen independently of r_w for all $w \neq v$.

Define the *natural clustering* of a graph $G \in \mathcal{G}$ by the classes $C_i' = C_i \cup \{v \in B \mid r_v = i\}$ ($i \in [k]$). We view N also as a graph formed by a disjoint union of the k cliques determined by $\{C_i'\}_{i \in [k]}$. This clustering will have a few disagreements because of the negative edges between different vertices $v, w \in B$ with $r_v = r_w$. For any pair of distinct elements $v, w \in B$, this happens with probability $1/k$. The cost of the optimal clustering of G is bounded by that of the natural clustering N, hence,

$$\mathbb{E}[\mathrm{OPT}] \leq \mathbb{E}[\mathrm{cost}(N)] = \frac{\binom{\alpha n}{2}}{k} \leq \frac{\alpha^2 n^2}{2k} = \frac{\varepsilon}{c} n^2.$$

We have to show that any algorithm making $< L/2$ queries to graphs drawn from $\mathcal{G}$ produces a clustering with expected cost larger than $2\varepsilon n^2$. Since all graphs in $\mathcal{G}$ induce the same subgraphs on A and B separately, we can assume without loss of generality that the algorithm queries only edges between A and B. Note that the neighborhoods in G of every pair of vertices from the same C_i are the same: $\Gamma_G(u) = \Gamma_G(v)$ for all $u, v \in C_i, i \in [k]$; moreover, u and v are joined by a positive edge. Therefore, if $u, v \in C_i$ but the algorithm assigns u and v to different clusters, either moving u to v's cluster or v to u's cluster will not decrease the cost. All in all, we can assume that the algorithm outputs k clusters $C_1', \ldots, C_k'$ with $C_i \subseteq C_i'$ for all i, plus (possibly) some clusters $C_{k+1}', \ldots, C_{k'}'$ ($k' \geq k$) involving only elements of B.

For $v \in B$, let $s_v \in [k']$ denote the cluster that the algorithm assigns v to. For every $v \in B$, let G_v denote the event that the algorithm queries (u, v) for some $u \in C_{r_v}$ and, whenever G_v does not hold, let us add a "fictitious" query to the algorithm between v and some arbitrary element of C_{s_v}. This ensures that whenever $r_v = s_v$, the last query of the algorithm verifies its guess and returns 1 if the correct cluster has been found. This adds at most $|B| \leq n \leq \frac{L}{2}$ queries in total. Let $Q_1, Q_2, \ldots, Q_z$ be the (random) sequence of queries issued by the algorithm and let $i_1^v, i_2^v, \ldots, i_{T_v}^v$ be the indices of those queries involving a fixed vertex $v \in B$. Note that r_v is independent of the response to all queries not involving v and, conditioned on the result of all queries up to time $t < i_{T_v}$, r_v is uniformly distributed among the set $\{i \in [k] \mid (Q_j \notin C_i \forall j < t)\}$, whose size is upper-bounded by $k - t + 1$. Therefore,

$$\Pr[Q_{i_t^v} \in C_{r_v} \mid Q_1, \ldots, Q_{i_t-1}] \leq \frac{1}{k - t + 1},$$

which becomes an equality if the algorithm does not query the same cluster twice. It follows by induction that

$$\Pr[\{Q_{i_1^v}, \dots, Q_{i_t^v}\} \cap C_{r_v} \neq \emptyset] \leq \frac{t}{k}. \tag{5.4}$$

Let M_v be the event that the algorithm makes more than $k/2$ queries involving v. The event $r_v = s_v$ is equivalent to G_v, i.e., the event $\{Q_{i_1^v}, \dots, Q_{i_{T_v}^v}\} \cap C_{r_v} \neq \emptyset$, because of our addition of one fictitious query. We have

$$\Pr[r_v = s_v] = \Pr[G_v] \leq \Pr[M_v] + \Pr[G_v \wedge \overline{M}_v].$$

In other words, either the algorithm makes many queries for v, or it hits the correct cluster with few queries. (Without fictitious queries, we would have to add a third term for the probability that the algorithm picks by chance the correct s_v.) We will use the first term $\Pr[M_v]$ to control the expected query complexity. The second term, $\Pr[G_v \wedge \overline{M}_v]$, is bounded by $\frac{1}{2}$ by (5.4) because $T_v \leq k/2$ whenever $\overline{M}_v$ holds. Hence,

$$\Pr[r_v \neq s_v] \geq \frac{1}{2} - \Pr[M_v],$$

so

$$\mathbb{E}[|\{v \in B \mid r_v \neq s_v\}|] = \sum_{v \in B} \Pr[r_v \neq s_v] \geq \frac{\alpha n}{2} - \left(\sum_{v \in B} \Pr[M_v] \right).$$

Each vertex $v \in B$ with $s_v \neq r_v$, causes disagreements with all of $C_{r_v} \subseteq C'_{r_v}$ and $C_{s_v} \subseteq C'_{r_v}$, introducing at least $2|A|/k \geq n/k$ new disagreements.

If we denote by X the cost of the clustering found and by Z the number of queries made, we have

$$\mathbb{E}[X] \geq \frac{n}{k} \mathbb{E}[|\{v \in B \mid r_v \neq s_v\}|]$$
$$\geq \frac{\alpha n^2}{2k} - \frac{n}{k} \left(\sum_{v \in B} \Pr[M_v] \right)$$
$$= 4\epsilon n^2 - \frac{n}{k} \left(\sum_{v \in B} \Pr[M_v] \right).$$

In particular, if $\mathbb{E}[X] \leq 2\epsilon n^2$, then we must have

$$\sum_{v \in B} \Pr[M_v] \geq \frac{2\epsilon n^2}{n/k} = 2\epsilon n k = \frac{n}{16c}.$$

But then we can lower bound the expected number of queries by

$$\mathbb{E}[Z] \geq \frac{k}{2} \sum_{v \in B} \Pr[M_v] \geq \frac{nk}{32c} = \frac{n}{1024c^2\epsilon} = \frac{n^3}{1024c^2T} = L,$$

of which at most $L/2$ are the fictitious queries we added. This completes the proof. □

Note that the above result also implies that any purely multiplicative approximation guarantee needs $\Omega(n^2)$ queries (e.g., by taking $T = 10n$).

5.1.2   A NON-ADAPTIVE ALGORITHM

There exist two main categories of query-efficient algorithms, *non-adaptive* and *adaptive*. The former choose their queries beforehand, while the latter can select the next query based on the response to previous queries. Thus, non-adaptive algorithms are better (can be run in parallel), while adaptive lower bounds are stronger. The QECC algorithm is an adaptive one. In fact, the queries made when picking a second pivot depend on the result of the queries made for the first pivot.

However, it is possible to modify QECC to make it non-adaptive. This can be achieved by observing that, instead of querying the second pivot based on the first one, it can be queried for the neighborhood of a random sample S of size $\frac{Q}{n-1}$. If we use the elements of S to find pivots, the same analysis shows that the output of this variant meets the exact same error bound of $3\text{OPT} + n^3/(2Q)$. The pseudocode for the adaptive variant of QECC is reported in Algorithm 5.20.

Algorithm 5.20 NA-QECC

Input: Unweighted graph $G = (V, E)$; query budget Q
Output: Clustering of G
Input: $G = (V, E)$; query budget Q
  $k \leftarrow \max\{t \leq n \mid (2n - 1 - t)t \leq 2Q\}$.
  Let $S = (v_1, \ldots, v_k)$ be a uniform random sample from V
  (with or without replacement)
  // Querying phase: find $\Gamma_G^+(v)$ for each $v \in S$
  for each $v \in S, w \in V, v < w$ **do**
    Query (v, w)
  end for
  // Clustering phase
  $R \leftarrow V$
  $i \leftarrow 1$
  while $i \leq k$ **do**
    if $v_i \in R$ **then**
      Output cluster $C = \{v_i\} \cup \Gamma_G^+(v_i) \cap R$.
      $R \leftarrow R \setminus C$
    end if
    $i \leftarrow i + 1$
  end while
  Output a separate singleton cluster for each remaining $v \in R$.

In practice, the adaptive QECC algorithm will run closer to the query budget, choosing more pivots and reducing the error somewhat below the theoretical bound, because it does not "waste" queries between a newly found pivot and the neighbors of previous pivots. Nevertheless, in settings where the similarity computations can be performed in parallel, it may become advantageous to use NA-QECC. Another benefit of the non-adaptive variant is that it gives a one-pass streaming algorithm for correlation clustering that uses only $\mathcal{O}(Q)$ space and processes edges in arbitrary order. These results are formally stated next.

Theorem 5.6 [García-Soriano et al., 2020]. *For any $Q > 0$, the NA-QECC algorithm (Algorithm 5.20) finds a clustering of G with expected cost at most $3\mathsf{OPT} + \frac{n^3}{2Q}$ making at most Q edge queries, and running in $\mathcal{O}(Q)$ time (under the assumption of unit-cost queries).*

Proof. The number of queries the algorithm makes is $S = (n-1) + (n-2) + \ldots (n-k) = \frac{2n-1-k}{2}k \leq Q$. Note that $\frac{n-1}{2}k \leq S \leq Q \leq (n-1)k$. The proof of the error bound proceeds exactly as in the proof of Theorem 5.2 (because $k \geq \frac{Q}{n-1}$). The running time of the querying phase of Non-adaptive QECC is $O(Q)$ and, assuming a hash table is use to store query answers, the expected running time of the second phase is bounded by $O(nk) = O(Q)$, because $k \leq \frac{2Q}{n-1}$.
$\square$

Another interesting consequence of this result (coupled with the lower bound presented above, in Theorem 5.5) is that adaptivity does not help for correlation clustering (beyond possibly a constant factor), in stark contrast to other problems where an exponential separation is known between the query complexity of adaptive and non-adaptive algorithms (e.g., Brody et al. [2011]).

5.1.3 A PRACTICAL IMPROVEMENT

An experimental study of algorithm QECC performed in García-Soriano et al. [2020] revealed that, while provably optimal up to constant factors, the algorithm sometimes returns solutions with poor recall of positive edges when the query budget is low. Intuitively, the reason is that, while picking a random pivot works in expectation, sometimes a low-degree pivot is chosen and all $|R| - 1$ queries are spent querying its neighbors, which may not be worth the effort for a small cluster when the query budget is tight. To entice the algorithm to choose higher-degree vertices (which would also improve the recall), García-Soriano et al. [2020] propose to bias it so that pivots are chosen with probability proportional to their positive degree within the subgraph induced by R. The conclusion of Lemma 5.4 remains unaltered in this case, but whether this change preserves the approximation guarantees from Ailon et al. [2008a] (which the overall guarantee of QECC rely on) remains open. García-Soriano et al. [2020] observe that this heuristic modification consistently improves recall and reduces the total number of disagreements in a variety of real-world datasets.

Such a heuristic variant of QECC is termed Heur-QECC and is outlined as Algorithm 5.21. A critical step of the heuristic is the computation of the degree of each vertex with a small number of queries. This cannot be afforded, but the following scheme is easily seen to choose each vertex $u \in R$ with probability $d_u/(2E)$, where d_u is the degree of u in the subgraph $G[R]$ induced by R and $E > 0$ is the total number of edges in $G[R]$:

1. Pick random pairs of vertices to query $(u, v) \in R \times R$ until an edge $(u, v) \in E$ is found.

2. Select the first endpoint u of this edge as a pivot.

When $E = 0$, this procedure will simply run out of queries to make.

Algorithm 5.21 Heur-QECC

Input: Unweighted graph $G = (V, E)$; query budget Q
Output: Clustering of G
  $R \leftarrow V$ // Unclustered vertices so far
  while $Q \geq |R| - 1$ **do**
    Pick a pair (u, v) from $R \times R$ uniformly at random.

    if $u \neq v$ **then**
      Query (u, v)
      $Q \leftarrow Q - 1$

      if $(u, v) \in E$ **then**
        Query all pairs (v, w) for $w \in R \setminus \{u, v\}$ to determine $\Gamma_G^+(v) \cap R$.
        $Q \leftarrow Q - |R| + 2$.
        Output cluster $C = \{v\} \cup \Gamma_G^+(v) \cap R$.
        $R \leftarrow R \setminus C$
      end if
    end if
  end while
  Output a separate singleton cluster for each remaining $v \in R$.

5.1.4   NEIGHBORHOOD QUERIES

A further interesting observation by García-Soriano et al. [2020] is that, if instead of pair similarity queries we allow *neighborhood oracles* (i.e., given v, we assume to obtain a linked list of all the *positive* neighbors of v in time linear in its length), then we can derive a constant-factor approximation algorithm to MIN-DISAGREE with $\mathcal{O}(n^{3/2})$ neighborhood queries, which can be significantly smaller than the number of edges in the graph. Indeed, Ailon and Liberty [2009] argue that with a neighborhood oracle, QwickCluster runs in time $\mathcal{O}(n + OPT)$; if $OPT \leq n^{3/2}$ this is $\mathcal{O}(n^{3/2})$. On the other hand, if $OPT > n^{3/2}$ we can stop the algorithm after $r = \sqrt{n}$ rounds, and by Lemma 5.4, we incur an additional cost of only $\mathcal{O}(n^{3/2}) = \mathcal{O}(OPT)$.

5.2   LOCAL CORRELATION CLUSTERING

In Bonchi et al. [2013a], the ideas behind the QECC algorithm presented on the previous section are taken a step forward to provide a *local* algorithm for correlation clustering. In *local correlation clustering* we are given the identifier of a single object and the goal is to return the label of the cluster to which it belongs in some globally consistent near-optimal clustering, using a small number of similarity queries. Local algorithms yield standard query-efficient algorithms for explicit clustering, but they are stronger because they enable us to answer the question "which cluster does this vertex belong to" quickly without clustering the whole graph. In particular, they also enable fast answering of same-cluster queries for any pair of vertices. Additionally, they imply (i) distributed and streaming clustering algorithms, (ii) fast estimators and testers for cluster edit distance, and (iii) property-preserving parallel reconstruction algorithms for clusterability in the model of Ailon et al. [2008b].

Specifically, Bonchi et al. [2013a] show that a simple modification of the QECC algorithm yields a local clustering algorithm. The idea is to find a set of $\mathcal{O}(1/\epsilon)$ pivots serving as cluster centers by running QECC on a random sample of $\mathcal{O}(1/\epsilon)$ vertices and then, given a new vertex v, select the *first* adjacent pivot as cluster label $\ell(v)$; if v is not adjacent to any pivot, then make it a singleton cluster. This algorithm attains a $(3, \varepsilon)$-approximation (a solution with cost at most $3\text{OPT} + \varepsilon n^2$, where OPT is the optimal cost) in time $\mathcal{O}(1/\varepsilon^2)$, independently of the graph size.

Bonchi et al. [2013a] also provide a second local algorithm achieving a fully additive $(1, \varepsilon)$-approximation (a solution with cost at most $3\text{OPT} + \varepsilon n^2$) with local query complexity $\text{poly}(1/\varepsilon)$ and time complexity $2^{\text{poly}(1/\varepsilon)}$. The explicit clustering can be found in time $n \cdot \text{poly}(1/\varepsilon) + 2^{\text{poly}(1/\varepsilon)}$. This is faster than the previously known PTAS for correlation clustering [Bansal et al., 2004, Giotis and Guruswami, 2006b]. This algorithm borrows ideas from the PTAS for dense MAX-CUT of Frieze et al. [2004], and uses low-rank approximations to the adjacency matrix of the graph via cut decompositions. Interestingly, while such approximations have been known for a long time, their implications for correlation clustering have been overlooked. Notably, implicit descriptions of these approximations are locally computable in time polynomial in the inverse of the approximation parameter. It is shown by Bonchi et al. [2013a] that in order to look for near-optimal clusterings, we can restrict the search to clusterings that "respect" a sufficiently fine weakly regular partition of the graph. Then they argue that this can be used to implicitly define a good approximate clustering: to cluster a given vertex, we first determine its piece in a regular partition, and then we look at which cluster contains this piece in the best coarsening of the partition.

5.3 LARGE-SCALE COMPUTING

Correlation clustering has been also addressed in computational models that go beyond the traditional, static one, i.e., models that are mainly devoted to large-scale computing. In the reminder we briefly overview the main results in this regard.

5.3.1 ONLINE SETTING

Mathieu et al. [2010] tackle the problem of *online correlation clustering*, that is correlation clustering in a context where vertices of a graph are supposed to arrive continuously, until the number $|V|$ of overall vertices has been reached ($|V|$ not known in advance). Upon arrival of a new vertex, an online algorithm updates the current clustering, by either creating a new singleton cluster, or adding the new vertex to a pre-existing cluster. It may also decide to merge some pre-existing clusters, whereas it is not allowed to split pre-existing clusters.

The main results of Mathieu et al. [2010] are as follows. For MIN-DISAGREE, they show that the natural greedy algorithm is $\mathcal{O}(|V|)$-competitive, and this is optimal up to a constant factor, even with randomization. The picture is better for MAX-AGREE, for which Mathieu et al. [2010] prove that the greedy algorithm is a $\frac{1}{2}$-competitive, but that no algorithm can be better than 0.803 competitive (0.834 for randomized algorithms). Moreover, they demonstrate that the optimal competitive ratio is better than $\frac{1}{2}$. This is accomplished with a non-greedy algorithm with competitive ratio $\frac{1}{2} + \varepsilon_0$, where ϵ_0 is a small absolute constant.

5.3.2 PARALLEL COMPUTING

Chierichetti et al. [2014] study correlation clustering in a *parallel* setting. Specifically, they focus on MIN-DISAGREE and develop a parallel version of the popular QwickCluster algorithm by Ailon et al. [2008a]. The main idea of Chierichetti et al. [2014]'s algorithm is to pick multiple pivots in parallel and repeat the process. While this idea is natural, Chierichetti et al. [2014] show that one has to be careful in how the pivots are selected in parallel, as a a less judicious choice can lead to poor-quality solutions. Chierichetti et al. [2014]'s algorithm runs in a logarithmic number of rounds for graphs with a constant number of positive neighbors, while for arbitrary graphs, the number of rounds gets further multiplied by the logarithm of the maximum positive degree. For complete graphs, it outputs a solution that is a close to 3-approximation to the optimal correlation-clustering solution. A noteworthy peculiarity of Chierichetti et al. [2014]'s algorithm is that it is very easy-to-implement in distributed (e.g., MapReduce), streaming, and message-passing models.

Other parallel algorithms for correlation clustering, specifically for MIN-DISAGREE, have been devised by Pan et al. [2015], namely C4 and ClusterWild. C4 is a parallel version of QwickCluster [Ailon et al., 2008a], achieving a 3-approximation in a poly-logarithmic number of rounds, by enforcing consistency between concurrently running peeling threads. Consistency in C4 is enforced using concurrency control, a notion extensively studied for databases transactions. ClusterWild is a coordination-free algorithm that waives consistency in favor of speed. The cost paid is an arbitrarily small loss in accuracy. ClusterWild is shown to achieve a $(3 + \varepsilon)\text{OPT} + \mathcal{O}(\varepsilon n \log^2 n)$ approximation (where OPT is optimal MIN-DISAGREE's objective-function value for the given input instance), in a poly-logarithmic number of rounds, with provable nearly-linear speedups.

Further work on correlation clustering in the parallel setting has been carried out by Cambus et al. [2021] and Cohen-Addad et al. [2021].

5.3.3 DYNAMIC DATA-STREAM MODEL

Ahn et al. [2015] focus on a *dynamic data-stream model*, where the stream consists of updates to the edge weights of a graph with $|V|$ vertices.

For MAX-AGREE, Ahn et al. [2015] devise two single-pass streaming algorithms needing $\tilde{\mathcal{O}}(|V|\varepsilon^{-2})$ space: (i) a polynomial-time $(1 - \varepsilon)$-approximation algorithm for bounded weights, and (ii) a $0.766(1 - \varepsilon)$-approximation algorithm for arbitrary weights, running in $\tilde{\mathcal{O}}(n\varepsilon^{-10})$ time. Both algorithms use cut sparsifiers as a data structure to estimate the number of agreements or disagreements of any clustering with a single-pass over the input stream. To obtain (i) using this data structure, they emulate the algorithm by Giotis and Guruswami [2006b], while to obtain (ii) a fast multiplicative weights update method is employed to approximately solve a semidefinite programming relaxation of MAX-AGREE, after which one can use the rounding scheme of Charikar et al. [2005].

As far as MIN-DISAGREE, Ahn et al. [2015] show that any algorithm that can test whether MIN-DISAGREE's objective function of a given stream is equal to zero in a single pass, for unit weights, must store $\Omega(|V|)$ bits. For arbitrary weights, the lower bound increases to $\Omega(|V| + |E^-|)$, where E^- is the set of negative edges. In this regard, Ahn et al. [2015] provide a single-pass algorithm that uses $s = \tilde{\mathcal{O}}(|V|\epsilon^{-2} + |E^-|)$ space and $\tilde{\mathcal{O}}(s^2)$ time, and achieves an $\mathcal{O}(\log|E^-|)$ approximation. The algorithm applies the multiplicative-weights update method to a sparsified version of a natural linear programming relaxation of MAX-AGREE. For unit weights when MIN-DISAGREE's objective function of a given stream is $\le t$, Ahn et al. [2015] provide a single-pass algorithm that uses $\tilde{\mathcal{O}}(|V| + t)$ space. The algorithm runs in polynomial time when t is constant. It uses a linear sketch to construct a spanning graph of the input graph in space $\tilde{\mathcal{O}}(|V|)$ and then builds all the forests formed by deleting at most t edges from it, which are then used to obtain a small collection of partitions among which the optimal solution belongs; bilinear sketches of the input graph are used to estimate the number of disagreements of each such partition.

Ahn et al. [2015] consider multiple-pass streaming algorithms as well. For unit weights, they present an $\mathcal{O}(\log\log|V|)$-pass algorithm that mimics the algorithm of Ailon et al. [2008a], and provides a 3-approximation with high probability in space $\mathcal{O}(n\text{polylog}n)$. Their analysis shows that $\mathcal{O}(\log\log n)$ rounds suffice to choose the pivots, whereas other approaches such as that of Chierichetti et al. [2014] require $\Omega(\log^2 n)$ rounds. For MIN-DISAGREE[k], on unit-weight graphs with $k \ge 3$, they give a $\min\{k - 1, \mathcal{O}(\log\log|V|)\}$-pass polynomial-time algorithm using $\tilde{\mathcal{O}}(|V|\varepsilon^{-2})$ space. This result is based on emulating an algorithm by Giotis and Guruswami [2006b] in the data-stream model.

CHAPTER 6

Conclusions and Open Problems

In this book we discussed *correlation clustering*, perhaps the most natural formulation of clustering. Given a set of objects and a pairwise similarity function between them, correlation clustering aims at partitioning the input objects in such a way to either minimize intra-cluster disagreements and inter-cluster agreements, or maximize intra-cluster agreements and inter-cluster disagreements, where (dis)agreements are measured in terms of the given similarity function. We have provided a comprehensive review of correlation clustering from several main perspectives, including fundamental theoretical results and algorithms, problem variants (constrained formulations, relaxed formulations, formulations handling graphs other than simple ones), advanced computational settings (large-scale, parallel, streaming, online), related problems, and applications.

Despite the considerable amount of work carried out so far, correlation clustering still comes with several open problems. In terms of fundamental results (Chapter 1), a classic open question is whether a combinatorial $\mathcal{O}(\log n)$-approximation algorithm for disagreement minimization on general graphs with arbitrary weights may exist (instead of the well-known LP solutions). Moreover, little appears to be known about the approximability of the weighted disagreement-minimization formulation where edge weights obey triangle inequality only, without probability constraint or other constraints involved. Despite this question was partially addressed by van Zuylen and Williamson [2007], an exhaustive answer is still missing, as they consider triangle-inequality-like hypotheses stronger than the standard triangle inequality. Other open problems along this line include further generalizing the extended weight bounds of Puleo and Milenkovic [2015] and going beyond the initial study of Mandaglio et al. [2021] on global weight bounds.

Another interesting direction for future work consists in investigating whether hard-constrained variants of correlation clustering (Chapter 2) may be tackled from a soft-constraint perspective. For instance, it might be worth studying correlation clustering with a fixed number of clusters where the bound on the number of output clusters has not to be necessarily met; rather, a penalty for its violation is considered in the objective function.

Similarly, it can be studied whether soft-fashion relaxations of correlation clustering (Chapter 3) may be rendered hard. An example in this regard is an alternative formulation of OVERLAPPING-CORRELATION-CLUSTERING that explicitates the maximum tolerated degree of overlap between clusters. Still in the context of constraints and relaxations, it might be interesting to investigate formulations involving multiple types of constraint/relaxation, rather than addressing each of them in isolation.

A recent variant of correlation clustering that has received not much attention so far is the cluster-wise variant of correlation clustering with local objectives (Section 3.2.2). It has been studied much less than its vertex-wise counterpart: one might consider extending the results derived for the latter to it.

As far as inputs going beyond the conventional ones (Chapter 4), to the best of our knowledge, correlation clustering has not been studied in special types of graphs such as, e.g., heterogeneous graphs, temporal graphs, or knowledge graphs.

Finally, for what concerns correlation clustering in special computational settings (Chapter 5), a number of interesting open questions arise in the context of query-efficient correlation clustering (Section 5.1). First, it could be investigated whether query-efficient algorithms based on the better LP-based correlation-clustering approximation algorithms can be used instead of the state-of-the-art ones devised by García-Soriano et al. [2020]. A further intriguing question is whether one can devise other graph-querying models that allow for improved theoretical results while being reasonable from a practical viewpoint.

Bibliography

A. Aboud. Correlation clustering with penalties and approximating the reordering buffer management problem. Ph.D. thesis, Computer Science Department, Technion, 2008. 74, 75

A. Agarwal, M. Charikar, K. Makarychev, and Y. Makarychev. $O(\sqrt{\log n})$ approximation algorithms for Min UnCut, Min 2CNF deletion, and directed cut problems. In *Proc. of ACM Symposium on Theory of Computing (STOC)*, pages 573–581, 2005. DOI: 10.1145/1060590.1060675 33

R. Agrawal, A. Halverson, K. Kenthapadi, N. Mishra, and P. Tsaparas. Generating labels from clicks. In *Proc. of ACM International Conference on Web Search and Data Mining (WSDM)*, pages 172–181, 2009. DOI: 10.1145/1498759.1498824 22

S. Ahmadi, S. Khuller, and B. Saha. Min-Max correlation clustering via MultiCut. In *Proc. of International Conference on Integer Programming and Combinatorial Optimization (IPCO)*, pages 13–26, 2019. DOI: 10.1007/978-3-030-17953-3_2 63, 74

S. Ahmadian, A. Epasto, R. Kumar, and M. Mahdian. Fair correlation clustering. In *Proc. of International Conference on Artificial Intelligence and Statistics (AISTATS)*, pages 4195–4205, 2020. 94, 95

K. J. Ahn, G. Cormode, S. Guha, A. McGregor, and A. Wirth. Correlation clustering in data streams. In *Proc. of International Conference on Machine Learning (ICML)*, pages 2237–2246, 2015. DOI: 10.1007/s00453-021-00816-9 111

N. Ailon and E. Liberty. Correlation clustering revisited: The true cost of error minimization problems. In *Proc. of International Colloquium on Automata, Languages, and Programming (ICALP)*, pages 24–36, 2009. DOI: 10.1007/978-3-642-02927-1_4 16, 17, 89, 108

N. Ailon, M. Charikar, and A. Newman. Aggregating inconsistent information: Ranking and clustering. *Journal of the ACM (JACM)*, 55:23:1–23:27, 2008a. DOI: 10.1145/1411509.1411513 5, 8, 9, 11, 12, 19, 20, 35, 41, 75, 78, 79, 82, 85, 86, 96, 100, 103, 107, 110, 111

N. Ailon, B. Chazelle, S. Comandur, and D. Liu. Property-preserving data reconstruction. *Algorithmica*, 51(2):160–182, 2008b. DOI: 10.1007/s00453-007-9075-9 109

N. Ailon, N. Avigdor-Elgrabli, E. Liberty, and A. van Zuylen. Improved approximation algorithms for bipartite correlation clustering. In *Proc. of European Symposium on Algorithms (ESA)*, pages 25–36, 2011. DOI: 10.1007/978-3-642-23719-5_3 77, 78, 79, 80

N. Ailon, N. Avigdor-Elgrabli, E. Liberty, and A. van Zuylen. Improved approximation algorithms for bipartite correlation clustering. *SIAM Journal on Computing (SICOMP)*, 41(5):1110–1121, 2012a. DOI: 10.1137/110848712 78, 79, 80

N. Ailon, R. Begleiter, and E. Ezra. Active learning using smooth relative regret approximations with applications. In *Proc. of Conference on Learning Theory (COLT)*, pages 19.1–19.20, 2012b. 99

K. Ambrosi. Aggregation binärer relationen in der qualitativen datenanalyse. *Metrika*, 31(1):274–274, 1984. DOI: 10.1007/bf01915211 2

N. Amit. The bicluster graph editing problem. Master's thesis, Tel Aviv University, 2004. 18, 64, 77, 78

Y. Anava, N. Avigdor-Elgrabli, and I. Gamzu. Improved theoretical and practical guarantees for chromatic correlation clustering. In *Proc. of World Wide Web Conference (WWW)*, pages 55–65, 2015. DOI: 10.1145/2736277.2741629 86, 87, 88, 90

C. E. Andrade, M. G. C. Resende, H. J. Karloff, and F. K. Miyazawa. Evolutionary algorithms for overlapping correlation clustering. In *Proc. of Genetic and Evolutionary Computation Conference (GECCO)*, pages 405–412, 2014. DOI: 10.1145/2576768.2598284 61

K. Andreev and H. Räcke. Balanced graph partitioning. *Theory of Computing Systems*, 39(6):929–939, 2006. DOI: 10.1007/s00224-006-1350-7 40

M. Asteris, A. Kyrillidis, D. Papailiopoulos, and A. Dimakis. Bipartite correlation clustering: Maximizing agreements. In *Artificial Intelligence and Statistics*, pages 121–129, 2016. 78

M. Balcan, A. Blum, and Y. Mansour. Improved equilibria via public service advertising. In C. Mathieu, Ed., *Proc. of ACM-SIAM Symposium on Discrete Algorithms (SODA)*, pages 728–737, 2009. DOI: 10.1137/1.9781611973068.80 72

N. Bansal, A. Blum, and S. Chawla. Correlation clustering. In *Proc. of IEEE Symposium on Foundations of Computer Science (FOCS)*, page 238, 2002. DOI: 10.1109/sfcs.2002.1181947 15

N. Bansal, A. Blum, and S. Chawla. Correlation clustering. *Machine Learning*, 56(1–3):89–113, 2004. DOI: 10.1023/b:mach.0000033116.57574.95 3, 4, 6, 8, 20, 23, 24, 28, 30, 35, 64, 82, 96, 97, 109

Y. Bartal. Probabilistic approximations of metric spaces and its algorithmic applications. In *Proc. of IEEE Symposium on Foundations of Computer Science (FOCS)*, pages 184–193, 1996. 72

J. P. Barthelemy and B. Monjardet. The median procedure in cluster analysis and social choice theory. *Mathematical Social Sciences*, 1(3):235–267, 1981. DOI: 10.1016/0165-4896(81)90041-x 2

A. Ben-Dor, R. Shamir, and Z. Yakhini. Clustering gene expression patterns. *Journal of Computational Biology*, 6(3/4):281–297, 1999. DOI: 10.1089/106652799318274 2, 3, 20

A. R. Benson, D. F. Gleich, and J. Leskovec. Higher-order organization of complex networks. *Science*, 353(6295):163–166, 2016. DOI: 10.1126/science.aad9029 95

A. Bhalgat, T. Chakraborty, and S. Khanna. Approximating pure Nash equilibrium in cut, party affiliation, and satisfiability games. In *Proc. of ACM Conference on Electronic Commerce (EC)*, pages 73–82, 2010. DOI: 10.1145/1807342.1807353 72, 73

A. Björn, H. K. Jörg, B. Thorsten, K. Ullrich, and A. H. Fred. Probabilistic image segmentation with closedness constraints. In *Proc. of IEEE International Conference on Computer Vision (ICCV)*, pages 2611–2618, 2011. DOI: 10.1109/iccv.2011.6126550 21

S. Böcker and J. Baumbach. Cluster editing. In *Proc. of Conference on Computability in Europe (CiE)*, pages 33–44, 2013. DOI: 10.1007/978-3-642-39053-1_5 18

S. Böcker, S. Briesemeister, Q. B. A. Bui, and A. Truß. Going weighted: Parameterized algorithms for cluster editing. *Theoretical Computer Science (TCS)*, 410(52):5467–5480, 2009. DOI: 10.1016/j.tcs.2009.05.006 18

F. Bonchi, A. Gionis, and A. Ukkonen. Overlapping correlation clustering. In *Proc. of IEEE International Conference on Data Mining (ICDM)*, pages 51–60, 2011. DOI: 10.1109/icdm.2011.114 51, 52, 53, 54, 55, 57, 58, 59, 60

F. Bonchi, A. Gionis, F. Gullo, and A. Ukkonen. Chromatic correlation clustering. In *Proc. of ACM SIGKDD International Conference on Knowledge Discovery and Data Mining*, pages 1321–1329, 2012. DOI: 10.1145/2339530.2339735 80, 81, 82, 83, 84, 85, 86, 87, 88, 90, 91

F. Bonchi, D. García-Soriano, and K. Kutzkov. Local correlation clustering. *CoRR*, 2013a. 15, 23, 99, 109

F. Bonchi, A. Gionis, and A. Ukkonen. Overlapping correlation clustering. *Knowledge and Information Systems (KAIS)*, 35(1):1–32, 2013b. DOI: 10.1007/s10115-012-0522-9 51, 52, 53, 54, 57, 61

F. Bonchi, D. García-Soriano, and E. Liberty. Correlation clustering: From theory to practice. In *Proc. of ACM SIGKDD International Conference on Knowledge Discovery and Data Mining*, page 1972, 2014a. DOI: 10.1145/2623330.2630808 xv, 2

F. Bonchi, A. Gionis, F. Gullo, and A. Ukkonen. Distance oracles in edge-labeled graphs. In *Proc. of International Conference on Extending Database Technology (EDBT)*, pages 547–558, 2014b. 80

F. Bonchi, A. Gionis, F. Gullo, C. E. Tsourakakis, and A. Ukkonen. Chromatic correlation clustering. *ACM Transactions on Knowledge Discovery from Data (TKDD)*, 9(4):34:1–34:24, 2015. DOI: 10.1145/2339530.2339735 2, 80, 81, 91, 92, 93

F. Bonchi, E. Galimberti, A. Gionis, B. Ordozgoiti, and G. Ruffo. Discovering polarized communities in signed networks. In *Proc. of International Conference on Information and Knowledge Management (CIKM)*, pages 961–970, 2019. DOI: 10.1145/3357384.3357977 21, 31, 75

J. C. Borda. Mémoire sur les élections au scrutin. *Histoire de l'Académie Royale des Sciences*, 1781. 19

S. Boyd and L. Vandenberghe. *Convex Optimization*. Cambridge University Press, 2004. DOI: 10.1017/cbo9780511804441 65

A. Z. Broder, M. Charikar, A. M. Frieze, and M. Mitzenmacher. Min-wise independent permutations. In *Proc. of ACM Symposium on Theory of Computing (STOC)*, pages 327–336, 1998. DOI: 10.1145/276698.276781 56

J. Brody, K. Matulef, and C. Wu. Lower bounds for testing computability by small width OBDDs. In *Proc. of International Conference on Theory and Applications of Models of Computation (TAMC)*, pages 320–331, 2011. DOI: 10.1007/978-3-642-20877-5_32 107

M. Cambus, D. Choo, H. Miikonen, and J. Uitto. Massively parallel correlation clustering in bounded arboricity graphs. In *Proc. of International Symposium on Distributed Computing (DISC)*, pages 15:1–15:18, 2021. 111

N. Cesa-Bianchi, C. Gentile, F. Vitale, and G. Zappella. A correlation clustering approach to link classification in signed networks. In *Proc. of Conference on Learning Theory (COLT)*, pages 34.1–34.20, 2012. 21

D. Chakrabarti, R. Kumar, and K. Punera. A graph-theoretic approach to webpage segmentation. In *Proc. of World Wide Web Conference (WWW)*, pages 377–386, 2008. DOI: 10.1145/1367497.1367549 22

M. Charikar, V. Guruswami, and A. Wirth. Clustering with qualitative information. In *Proc. of IEEE Symposium on Foundations of Computer Science (FOCS)*, pages 524–533, 2003. DOI: 10.1109/sfcs.2003.1238225 77, 87

M. Charikar, V. Guruswami, and A. Wirth. Clustering with qualitative information. *Journal of Computer and System Sciences (JCSS)*, 71(3):360–383, 2005. DOI: 10.1016/j.jcss.2004.10.012 8, 13, 15, 35, 36, 37, 38, 39, 65, 66, 71, 72, 78, 87, 96, 111

M. Charikar, N. Gupta, and R. Schwartz. Local guarantees in graph cuts and clustering. In *Proc. of International Conference on Integer Programming and Combinatorial Optimization (IPCO)*, pages 136–147, 2017. DOI: 10.1007/978-3-319-59250-3_12 67, 69, 70, 71, 72, 73

S. Chawla, R. Krauthgamer, R. Kumar, Y. Rabani, and D. Sivakumar. On the hardness of approximating Multicut and Sparsest-Cut. *Computational Complexity*, 15(2):94–114, 2006. DOI: 10.1007/s00037-006-0210-9 14, 35

S. Chawla, K. Makarychev, T. Schramm, and G. Yaroslavtsev. Near optimal LP rounding algorithm for correlation clustering on complete and complete k-partite graphs. In *Proc. of ACM Symposium on Theory of Computing (STOC)*, pages 219–228, 2015. DOI: 10.1145/2746539.2746604 12, 13, 78

Y. Chen, A. Jalali, S. Sanghavi, and H. Xu. Clustering partially observed graphs via convex optimization. *Journal of Machine Learning Research (JMLR)*, 15(1):2213–2238, 2014. 97

F. Chierichetti, R. Kumar, S. Pandey, and S. Vassilvitskii. Finding the Jaccard median. In *Proc. of ACM-SIAM Symposium on Discrete Algorithms (SODA)*, pages 293–311, 2010. DOI: 10.1137/1.9781611973075.25 59

F. Chierichetti, N. N. Dalvi, and R. Kumar. Correlation clustering in MapReduce. In *Proc. of ACM SIGKDD International Conference on Knowledge Discovery and Data Mining*, pages 641–650, 2014. DOI: 10.1145/2623330.2623743 110, 111

F. Chierichetti, R. Kumar, S. Lattanzi, and S. Vassilvitskii. Fair clustering through fairlets. In *Proc. of Conference on Advances in Neural Information Processing Systems (NeurIPS)*, pages 5029–5037, 2017. 94

G. Christodoulou, V. S. Mirrokni, and A. Sidiropoulos. Convergence and approximation in potential games. *Theoretical Computer Science (TCS)*, 438:13–27, 2012. DOI: 10.1016/j.tcs.2012.02.033 72

M. Chrobak, C. Dürr, A. Fabijan, and B. J. Nilsson. Online clique clustering. *Algorithmica*, 82(4):938–965, 2020. DOI: 10.1007/s00453-019-00625-1 18

W. W. Cohen and J. Richman. Learning to match and cluster entity names. In *ACM SIGIR Workshop on Mathematical/Formal Methods in Information Retrieval*, 2001. 3

W. W. Cohen and J. Richman. Learning to match and cluster large high-dimensional data sets for data integration. In *Proc. of ACM SIGKDD International Conference on Knowledge Discovery and Data Mining*, pages 475–480, 2002. DOI: 10.1145/775047.775116 3, 20

V. Cohen-Addad, S. Lattanzi, S. Mitrovic, A. Norouzi-Fard, N. Parotsidis, and J. Tarnawski. Correlation clustering in constant many parallel rounds. In *Proc. of International Conference on Machine Learning (ICML)*, pages 2069–2078, 2021. 111

T. Coleman, J. Saunderson, and A. Wirth. A local-search 2-approximation for 2-correlation-clustering. In *Proc. of European Symposium on Algorithms (ESA)*, pages 308–319, 2008a. DOI: 10.1007/978-3-540-87744-8_26 23, 30, 31, 32, 33

T. Coleman, J. Saunderson, and A. Wirth. Spectral clustering with inconsistent advice. In *Proc. of International Conference on Machine Learning (ICML)*, pages 152–159, 2008b. DOI: 10.1145/1390156.1390176 33, 34

J.-A.-N. C. D. Condorcet. *Éssai sur l'application de l'analyse á la probabilité des décisions rendues á la pluralité des voix*, 1785. 19

T. H. Cormen, C. E. Leiserson, R. L. Rivest, and C. Stein. *Introduction to Algorithms*. MIT Press, 2009. DOI: 10.2307/2583667 xiv

C. Crespelle, P. G. Drange, F. V. Fomin, and P. A. Golovach. A survey of parameterized algorithms and the complexity of edge modification. *CoRR*, 2020. 18

I. Csiszar and G. Tusnady. Information geometry and alternating minimization procedures. *Statistics and Decisions*, 1984. 90

B. DasGupta, G. A. Enciso, E. D. Sontag, and Y. Zhang. Algorithmic and complexity results for decompositions of biological networks into monotone subsystems. *Biosystems*, 90(1):161–178, 2007. DOI: 10.1016/j.biosystems.2006.08.001 20

H. Dau and O. Milenkovic. Latent network features and overlapping community discovery via boolean intersection representations. *IEEE/ACM Transactions on Networking*, 25(5):3219–3234, 2017. DOI: 10.1109/tnet.2017.2728638 57

W. F. de la Vega and C. Kenyon. A randomized approximation scheme for Metric MAX-CUT. In *Proc. of IEEE Symposium on Foundations of Computer Science (FOCS)*, pages 468–471, 1998. DOI: 10.1109/sfcs.1998.743497 15, 29

J. Dean and S. Ghemawat. MapReduce: Simplified data processing on large clusters. *Communications of the ACM (CACM)*, 51(1):107–113, 2008. DOI: 10.1145/1327452.1327492 61

E. D. Demaine, D. Emanuel, A. Fiat, and N. Immorlica. Correlation clustering in general weighted graphs. *Theoretical Computer Science (TCS)*, 361(2–3):172–187, 2006. DOI: 10.1016/j.tcs.2006.05.008 2, 13, 14, 35, 44, 45, 71, 72, 74, 77, 87

A. Dessmark, J. Jansson, A. Lingas, E.-M. Lundell, and M. Persson. On the approximability of maximum and minimum edge clique partition problems. *International Journal of Foundations of Computer Science (IJFCS)*, 18(02):217–226, 2007. DOI: 10.1142/s0129054107004656 18

Devvrit, R. Krishnaswamy, and N. Rajaraman. Robust correlation clustering. In *Approximation, Randomization, and Combinatorial Optimization. Algorithms and Techniques (APPROX/RANDOM)*, pages 33:1–33:18, 2019. 75

M. E. Dickison, M. Magnani, and L. Rossi. *Multilayer Social Networks*. Cambridge University Press, 2016. DOI: 10.1017/cbo9781139941907 91

C. Ding, X. He, and H. D. Simon. On the equivalence of nonnegative matrix factorization and spectral clustering. In *Proc. of SIAM International Conference on Data Mining (SDM)*, pages 606–610, 2005. DOI: 10.1137/1.9781611972757.70 56

C. Dwork, R. Kumar, M. Naor, and D. Sivakumar. Rank aggregation methods for the Web. In *Proc. of World Wide Web Conference (WWW)*, pages 613–622, 2001. DOI: 10.1145/371920.372165 22

M. Elsner and E. Charniak. You talking to me? A corpus and algorithm for conversation disentanglement. In *Proc. of Annual Meeting of the Association for Computational Linguistics (ACL)*, pages 834–842, 2008. 20

P. Erdös, A. W. Goodman, and L. Pósa. The representation of a graph by set intersections. *Canadian Journal of Mathematics*, 18:106–112, 1966. DOI: 10.4153/cjm-1966-014-3 55

A. Fabrikant, C. H. Papadimitriou, and K. Talwar. The complexity of pure Nash equilibria. In *Proc. of ACM Symposium on Theory of Computing (STOC)*, pages 604–612, 2004. DOI: 10.1145/1007352.1007445 72

J. Fakcharoenphol, S. Rao, and K. Talwar. A tight bound on approximating arbitrary metrics by tree metrics. *Journal of Computer and System Sciences (JCSS)*, 69(3):485–497, 2004. DOI: 10.1016/j.jcss.2004.04.011 72

W. Fan, J. Li, S. Ma, N. Tang, and Y. Wu. Adding regular expressions to graph reachability and pattern queries. In *Proc. IEEE International Conference on Data Engineering (ICDE)*, pages 39–50, 2011. DOI: 10.1109/icde.2011.5767858 80

X. Z. Fern and C. E. Brodley. Solving cluster ensemble problems by bipartite graph partitioning. In *Proc. of International Conference on Machine Learning (ICML)*, 2004. DOI: 10.1145/1015330.1015414 77

A. Figueroa, A. Goldstein, T. Jiang, M. Kurowski, A. Lingas, and M. Persson. Approximate clustering of fingerprint vectors with missing values. In *Proc. of Australasian Theory Symposium (CATS)*, pages 57–60, 2005. 18

J. R. Finkel and C. D. Manning. Enforcing transitivity in coreference resolution. In *Proc. of Annual Meeting of the Association for Computational Linguistics (ACL)*, pages 45–48, 2008. DOI: 10.3115/1557690.1557703 20

A. M. Frieze, R. Kannan, and S. Vempala. Fast Monte Carlo algorithms for finding low-rank approximations. *Journal of the ACM (JACM)*, 51(6):1025–1041, 2004. DOI: 10.1145/1039488.1039494 109

T. Fukunaga. LP-based pivoting algorithm for higher-order correlation clustering. *Computing and Combinatorics*, pages 51–62, 2018. DOI: 10.1007/978-3-319-94776-1_5 96

E. Galimberti, F. Bonchi, and F. Gullo. Core decomposition and densest subgraph in multi-layer networks. In *Proc. of International Conference on Information and Knowledge Management (CIKM)*, pages 1807–1816, 2017. DOI: 10.1145/3132847.3132993 91

E. Galimberti, F. Bonchi, F. Gullo, and T. Lanciano. Core decomposition in multilayer net-works: Theory, algorithms, and applications. *ACM Transactions on Knowledge Discovery from Data (TKDD)*, 14(1):11:1–11:40, 2020. DOI: 10.1145/3369872 91

D. García-Soriano, K. Kutzkov, F. Bonchi, and C. E. Tsourakakis. Query-efficient correlation clustering. In *Proc. of World Wide Web Conference (WWW)*, pages 1468–1478, 2020. DOI: 10.1145/3366423.3380220 99, 100, 101, 103, 107, 108, 114

M. R. Garey and D. S. Johnson. *Computers and Intractability: A Guide to the Theory of NP-Completeness*. W. H. Freeman & Co., 1979. xiv, 4, 53, 64

M. R. Garey, D. S. Johnson, and L. J. Stockmeyer. Some simplified NP-complete graph problems. *Theoretical Computer Science (TCS)*, 1(3):237–267, 1976. DOI: 10.1016/0304-3975(76)90059-1 25

N. Garg, V. V. Vazirani, and M. Yannakakis. Approximate Max-Flow Min-(Multi)Cut theo-rems and their applications. *SIAM Journal on Computing (SICOMP)*, 25(2):235–251, 1996. DOI: 10.1137/s0097539793243016 14, 46, 47, 49, 72

F. Geerts and R. Ndindi. Bounded correlation clustering. *International Journal of Data Science and Analytics (IJDSA)*, 1(1):17–35, 2016. DOI: 10.1007/s41060-016-0005-2 42, 43, 44, 45, 46, 47, 48, 49

A. Gionis, H. Mannila, and P. Tsaparas. Clustering aggregation. *ACM Transactions on Knowl-edge Discovery from Data (TKDD)*, 1(1):4, 2007. DOI: 10.1145/1217299.1217303 2, 18, 19, 22, 35

I. Giotis and V. Guruswami. Correlation clustering with a fixed number of clusters. In *Proc. of ACM-SIAM Symposium on Discrete Algorithms (SODA)*, pages 1167–1176, 2006a. DOI: 10.1145/1109557.1109686 15

I. Giotis and V. Guruswami. Correlation clustering with a fixed number of clusters. *Theory of Computing*, 2(13):249–266, 2006b. DOI: 10.1145/1109557.1109686 23, 25, 26, 27, 28, 29, 30, 32, 33, 109, 111, 112

D. F. Gleich, N. Veldt, and A. Wirth. Correlation clustering generalized. In *Proc. of International Symposium on Algorithms and Computation (ISAAC)*, pages 44:1–44:13, 2018. 96

M. X. Goemans and D. P. Williamson. Improved approximation algorithms for Maximum Cut and Satisfiability problems using semidefinite programming. *Journal of the ACM (JACM)*, 42(6):1115–1145, 1995. DOI: 10.1145/227683.227684 15, 32

O. Goldreich, S. Goldwasser, and D. Ron. Property testing and its connection to learning and approximation. *Journal of the ACM (JACM)*, 45(4):653–750, 1998. DOI: 10.1145/285055.285060 27

M. Grötschel and Y. Wakabayashi. A cutting plane algorithm for a clustering problem. *Mathematical Programming*, 45(1–3):59–96, 1989. DOI: 10.1007/bf01589097 3

M. Grötschel and Y. Wakabayashi. Facets of the clique partitioning polytope. *Mathematical Programming*, 47(1–3):367–387, 1990. DOI: 10.1007/bf01580870 3

J. Guo, J. Gramm, F. Hüffner, R. Niedermeier, and S. Wernicke. Compression-based fixed-parameter algorithms for feedback vertex set and edge bipartization. *Journal of Computer and System Sciences (JCSS)*, 72(8):1386–1396, 2006. DOI: 10.1016/j.jcss.2006.02.001 34

J. Guo, F. Hüffner, C. Komusiewicz, and Y. Zhang. Improved algorithms for bicluster editing. In *Proc. of International Conference on Theory and Applications of Models of Computation (TAMC)*, pages 445–456, 2008. DOI: 10.1007/978-3-540-79228-4_39 78

A. Gupta, R. Krauthgamer, and J. R. Lee. Bounded geometries, fractals, and low-distortion embeddings. In *Proc. of IEEE Symposium on Foundations of Computer Science (FOCS)*, pages 534–543, 2003. DOI: 10.1109/sfcs.2003.1238226 72

O. Hassanzadeh, F. Chiang, R. J. Miller, and H. C. Lee. Framework for evaluating clustering algorithms in duplicate detection. *Proc. of the VLDB Endowment (PVLDB)*, 2(1):1282–1293, 2009. DOI: 10.14778/1687627.1687771 2, 20

Z. He, S. Xie, R. Zdunek, G. Zhou, and A. Cichocki. Symmetric nonnegative matrix factorization: Algorithms and applications to probabilistic clustering. *IEEE Transactions on Neural Networks*, 22(12):2117–2131, 2011. DOI: 10.1109/tnn.2011.2172457 56

P. Hell and D. G. Kirkpatrick. Algorithms for degree constrained graph factors of minimum deficiency. *Journal of Algorithms*, 14(1):115–138, 1993. DOI: 10.1006/jagm.1993.1006 41

T. C. Hu. Multi-commodity network flows. *Operational Research*, 11(3):344–360, 1963. DOI: 10.1287/opre.11.3.344 43

F. Hüffner, N. Betzler, and R. Niedermeier. Optimal edge deletions for signed graph balancing. In *Proc. of International Workshop on Experimental Algorithms (WEA)*, pages 297–310, 2007. DOI: 10.1007/978-3-540-72845-0_23 32, 34

V. Il'Ev and A. Navrotskaya. A local search for a graph correlation clustering. In *Supplementary Proc. of International Conference on Discrete Optimization and Operations Research and Scientific School (DOOR)*, pages 510–515, 2016. DOI: 10.1063/1.4965325 23, 31

P. Indyk. A sublinear time approximation scheme for clustering in metric spaces. In *Proc. of IEEE Symposium on Foundations of Computer Science (FOCS)*, pages 154–159, 1999. DOI: 10.1109/sffcs.1999.814587 29

P. Indyk and R. Motwani. Approximate nearest neighbors: Towards removing the curse of dimensionality. In *Proc. of ACM Symposium on Theory of Computing (STOC)*, pages 604–613, 1998. DOI: 10.1145/276698.276876 56

S. Ji, D. Xu, M. Li, and Y. Wang. Approximation algorithms for two variants of correlation clustering problem. *Journal of Combinatorial Optimization*, 2020. DOI: 10.1007/s10878-020-00612-1 42

R. Jin, H. Hong, H. Wang, N. Ruan, and Y. Xiang. Computing label-constraint reachability in graph databases. In *Proc. ACM SIGMOD International Conference on Management of Data*, pages 123–134, 2010. DOI: 10.1145/1807167.1807183 80

T. Joachims and J. E. Hopcroft. Error bounds for correlation clustering. In *Proc. of International Conference on Machine Learning (ICML)*, pages 385–392, ACM, 2005. DOI: 10.1145/1102351.1102400 89, 97

D. S. Johnson, C. H. Papadimitriou, and M. Yannakakis. How easy is local search? *Journal of Computer and System Sciences (JCSS)*, 37(1):79–100, 1988. DOI: 10.1016/0022-0000(88)90046-3 73

D. V. Kalashnikov, Z. Chen, S. Mehrotra, and R. Nuray-Turan. Web people search via connection analysis. *IEEE Transactions on Knowledge and Data Engineering (TKDE)*, 20(11):1550–1565, 2008. DOI: 10.1109/tkde.2008.78 22

S. Kalhan, K. Makarychev, and T. Zhou. Correlation clustering with local objectives. In *Proc. of Conference on Advances in Neural Information Processing Systems (NeurIPS)*, pages 9341–9350, 2019. 67, 68, 70, 71, 72, 74

N. Karmarkar and K. G. Ramakrishnan. Computational results of an interior point algorithm for large scale linear programming. *Mathematical Programming*, 52:555–586, 1991. DOI: 10.1007/bf01582905 38

M. Karpinski and W. Schudy. Linear time approximation schemes for the Gale–Berlekamp game and related minimization problems. In *Proc. of ACM Symposium on Theory of Computing (STOC)*, pages 313–322, 2009. DOI: 10.1145/1536414.1536458 23, 26, 28, 29, 30

J. G. Kemeny. Mathematics without numbers. *Daedalus*, 88:571–591, 1959. 19

A. Khan, F. Gullo, T. Wohler, and F. Bonchi. Top-k reliable edge colors in uncertain graphs. In *Proc. of International Conference on Information and Knowledge Management (CIKM)*, pages 1851–1854, 2015. DOI: 10.1145/2806416.2806619 80

A. Khan, F. Bonchi, F. Gullo, and A. Nufer. Conditional reliability in uncertain graphs. *IEEE Transactions on Knowledge and Data Engineering (TKDE)*, 30(11):2078–2092, 2018. DOI: 10.1109/tkde.2018.2816653 80

S. Khot. On the power of unique 2-prover 1-round games. In *Proc. of ACM Symposium on Theory of Computing (STOC)*, pages 767–775, 2002. DOI: 10.1145/509907.510017 14, 33, 35

S. Khot, G. Kindler, E. Mossel, and R. O'Donnell. Optimal inapproximability results for MAX-CUT and other 2-variable CSPs. *SIAM Journal on Computing (SICOMP)*, 37(1):319–357, 2007. DOI: 10.1137/s0097539705447372 33

S. Kim, S. Nowozin, P. Kohli, and C. D. Yoo. Higher-order correlation clustering for image segmentation. In *Proc. of Conference on Advances in Neural Information Processing Systems (NeurIPS)*, pages 1530–1538, 2011. 2, 21, 95

J. Kleinberg and E. Tardos. *Algorithm Design*. Pearson Education India, 2006. xiv

N. Klodt, L. Seifert, A. Zahn, K. Casel, D. Issac, and T. Friedrich. A color-blind 3-approximation for chromatic correlation clustering and improved heuristics. In *Proc. of ACM SIGKDD International Conference on Knowledge Discovery and Data Mining*, pages 882–891, 2021. DOI: 10.1145/3447548.3467446 87, 90

M. Křivánek and J. Morávek. Np-hard problems in hierarchical-tree clustering. *Acta Informatica*, 23(3):311–323, 1986. DOI: 10.1007/bf00289116 3

S. Kushagra, S. Ben-David, and I. Ilyas. Semi-supervised clustering for de-duplication. In *Proc. of International Conference on Artificial Intelligence and Statistics (AISTATS)*, pages 1659–1667, 2019. 2

D. D. Lee and H. S. Seung. Algorithms for non-negative matrix factorization. In *Proc. of Conference on Advances in Neural Information Processing Systems (NeurIPS)*, pages 556–562, 2001. 56

K.-M. Lee, B. Min, and K.-I. Goh. Towards real-world complexity: An introduction to multiplex networks. *The European Physical Journal B*, 88(2), 2015. DOI: 10.1140/epjb/e2015-50742-1 91

P. Li, H. Dau, G. J. Puleo, and O. Milenkovic. Motif clustering and overlapping clustering for social network analysis. In *Proc. of IEEE Conference on Computer Communications (INFO-COM)*, pages 1–9, 2017. DOI: 10.1109/infocom.2017.8056956 55, 95, 96

C. Lin, Y. rae Cho, W. chang Hwang, P. Pei, and A. Zhang. Clustering methods in protein-protein interaction networks. In X. Hu and Y. Pan, Eds., *Knowledge Discovery in Bioinformatics: Techniques, Methods, and Application*. Wiley, 2007. DOI: 10.1002/9780470124642.ch16 80

L. Lovasz. Covering and coloring of hypergraphs. In *Proc. of Sourtheastern International Conference on Combinatorics, Graph Theory and Computing (SEICCGTC)*, pages 3–12, 1973. 24

S. C. Madeira and A. L. Oliveira. Biclustering algorithms for biological data analysis: A survey. *IEEE/ACM Transactions on Computational Biology and Bioinformatics*, 1(1), 2004. DOI: 10.1109/tcbb.2004.2 77

K. Makarychev, Y. Makarychev, and A. Vijayaraghavan. Correlation clustering with noisy partial information. In *Proc. of Conference on Learning Theory (COLT)*, pages 1321–1342, 2015. 97, 98

U. Manber. *Introduction to Algorithms: A Creative Approach*. Addison-Wesley Longman Publishing Co., Inc., 1989. xiv

D. Mandaglio, A. Tagarelli, and F. Gullo. In and out: Optimizing overall interaction in probabilistic graphs under clustering constraints. In *Proc. of ACM SIGKDD International Conference on Knowledge Discovery and Data Mining*, pages 1371–1381, 2020. DOI: 10.1145/3394486.3403190 21

D. Mandaglio, A. Tagarelli, and F. Gullo. Correlation clustering with global weight bounds. In *Proc. of European Machine Learning and Principles and Practice of Knowledge Discovery in Databases (ECML PKDD)*, pages 499–515, 2021. DOI: 10.1007/978-3-030-86520-7_31 11, 113

F. Marcotorchino and P. Michaud. Optimization in exploratory data analysis. In *Proc. International Symposium on Operations Research*, 1981a. 2

J. Marcotorchino and P. Michaud. Heuristic approach of the similarity aggregation problem. *Methods of Operations Research*, 43:395–404, 1981b. 2

C. Mathieu and W. Schudy. Correlation clustering with noisy input. In *Proc. of ACM-SIAM Symposium on Discrete Algorithms (SODA)*, pages 712–728, 2010. DOI: 10.1137/1.9781611973075.58 89, 97

C. Mathieu, O. Sankur, and W. Schudy. Online correlation clustering. In *Proc. International Symposium on Theoretical Aspects of Computer Science (STACS)*, pages 573–584, 2010. 110

A. McCallum and B. Wellner. Conditional models of identity uncertainty with application to noun coreference. In *Proc. of Conference on Advances in Neural Information Processing Systems (NeurIPS)*, pages 905–912, 2005. 2, 22

S. Mehrotra. On the implementation of a primal-dual interior point method. *SIAM Journal on Optimization*, 2(4):575–601, 1992. DOI: 10.1137/0802028 38

P. Miettinen. On the positive-negative partial set cover problem. *Information Processing Letters*, 108(4), 2008. DOI: 10.1016/j.ipl.2008.05.007 60

B. Mirkin. The problems of approximation in space of relations and qualitative data analysis. *Information and Remote Control*, 35:1424–1431, 1974. 2

B. G. Mirkin. *Group Choice*. Halsted Press, 1979. 2

A. Miyauchi, T. Sonobe, and N. Sukegawa. Exact clustering via integer programming and maximum satisfiability. In *32nd AAAI Conference on Artificial Intelligence*, 2018. 18

D. Monderer and L. S. Shapley. Potential games. *Games and Economic Behavior*, 14(1):124–143, 1996. DOI: 10.1006/game.1996.0044 72

M. E. Newman. Finding community structure in networks using the eigenvectors of matrices. *Physical Review E*, 74(3):036104, 2006. DOI: 10.1103/physreve.74.036104 35

O. Opitz and M. Schader. Analyse qualitativer daten: Einführung und übersicht. *Operations-Research-Spektrum*, 6(2):67–83, 1984. DOI: 10.1007/bf01721080 2

X. Pan, D. S. Papailiopoulos, S. Oymak, B. Recht, K. Ramchandran, and M. I. Jordan. Parallel correlation clustering on big graphs. In *Proc. of Conference on Advances in Neural Information Processing Systems (NeurIPS)*, pages 82–90, 2015. 110

D. Pandove, S. Goel, and R. Rani. Correlation clustering methodologies and their fundamental results. *Expert Systems*, 35(1), 2018. DOI: 10.1111/exsy.12229 22

M. Pilipczuk, M. Pilipczuk, and M. Wrochna. Edge bipartization faster than 2^k. *Algorithmica*, 81(3):917–966, 2019. DOI: 10.1007/s00453-017-0319-z 34

J. Pouget-Abadie, K. Aydin, W. Schudy, K. Brodersen, and V. Mirrokni. Variance reduction in bipartite experiments through correlation clustering. In *Proc. of Conference on Advances in Neural Information Processing Systems (NeurIPS)*, pages 13309–13319, 2019. 2

G. J. Puleo and O. Milenkovic. Correlation clustering with constrained cluster sizes and extended weights bounds. *SIAM Journal on Optimization*, 25(3):1857–1872, 2015. DOI: 10.1137/140994198 11, 34, 35, 36, 37, 38, 39, 40, 41, 42, 113

G. J. Puleo and O. Milenkovic. Correlation clustering and biclustering with locally bounded errors. *IEEE Transactions on Information Theory*, 64(6):4105–4119, 2018. DOI: 10.1109/tit.2018.2819696 61, 62, 63, 64, 65, 66, 67, 68, 69

A. P. Punnen and R. Zhang. Analysis of an approximate greedy algorithm for the maximum edge clique partitioning problem. *Discrete Optimization*, 9(3):205–208, 2012. DOI: 10.1016/j.disopt.2012.05.002 18

A. Ramachandran, N. Feamster, and S. Vempala. Filtering spam with behavioral blacklisting. In *Proc. of ACM Conference on Computer and Communications Security (CCS)*, pages 342–351, 2007. DOI: 10.1145/1315245.1315288 2

S. Rao. Small distortion and volume preserving embeddings for planar and euclidean metrics. In *Proc. of International Symposium on Computational Geometry (SoCG)*, pages 300–306, 1999. DOI: 10.1145/304893.304983 72

S. Régnier. Sur quelques aspects mathématiques des problèmes de classification automatique. *ICC Bulletin*, 4(3):175–191, 1965. 2

A. A. Schäffer and M. Yannakakis. Simple local search problems that are hard to solve. *SIAM Journal on Computing (SICOMP)*, 20(1):56–87, 1991. 72

E. R. Scheinerman and K. Tucker. Modeling graphs using dot product representations. *Computational Statistics*, 25(1), 2010. DOI: 10.1007/s00180-009-0158-8 57

R. Shamir, R. Sharan, and D. Tsur. Cluster graph modification problems. *Discrete Applied Mathematics*, 144(1–2):173–182, 2004. DOI: 10.1016/j.dam.2004.01.007 3, 17, 18, 24, 25

N. Sukegawa and A. Miyauchi. A note on the complexity of the maximum edge clique partitioning problem with respect to the clique number. *Discrete Optimization*, 10(4):331–332, 2013. DOI: 10.1016/j.disopt.2013.08.004 18

C. Swamy. Correlation clustering: Maximizing agreements via semidefinite programming. In *Proc. of ACM-SIAM Symposium on Discrete Algorithms (SODA)*, pages 526–527, 2004. 15, 33, 77

A. Tagarelli, A. Amelio, and F. Gullo. Ensemble-based community detection in multi-layer networks. *Data Mining and Knowledge Discovery*, 31(5):1506–1543, 2017. DOI: 10.1007/s10618-017-0528-8 91

C. E. Tsourakakis. Provably fast inference of latent features from networks: With applications to learning social circles and multilabel classification. In *Proc. of World Wide Web Conference (WWW)*, pages 1111–1121, 2015. DOI: 10.1145/2736277.2741128 55, 57

J. M. M. van Rooij, M. E. van Kooten Niekerk, and H. L. Bodlaender. Partition into triangles on bounded degree graphs. *Theory of Computing Systems (TOCS)*, 52(4):687–718, 2013. DOI: 10.1007/s00224-012-9412-5 63

A. van Zuylen and D. P. Williamson. Deterministic algorithms for rank aggregation and other ranking and clustering problems. In *Proc. of International Workshop on Approximation and Online Algorithms (WAOA)*, pages 260–273, 2007. DOI: 10.1007/978-3-540-77918-6_21 11, 113

A. van Zuylen and D. P. Williamson. Deterministic pivoting algorithms for constrained ranking and clustering problems. *Mathematics of Operations Research*, 34(3):594–620, 2009. DOI: 10.1287/moor.1090.0385 11, 12

V. V. Vazirani. *Approximation Algorithms*. Springer-Verlag, Berlin, Heidelberg, 2001. DOI: 10.1007/978-3-662-04565-7 xiv, 46

N. Veldt, D. F. Gleich, and A. Wirth. A correlation clustering framework for community detection. In *Proc. of World Wide Web Conference (WWW)*, pages 439–448, 2018. DOI: 10.1145/3178876.3186110 21

N. Veldt, A. Wirth, and D. F. Gleich. Parameterized correlation clustering in hypergraphs and bipartite graphs. In *Proc. of ACM SIGKDD International Conference on Knowledge Discovery and Data Mining*, pages 1868–1876, 2020. DOI: 10.1145/3394486.3403238 78

M. Vlachos, F. Fusco, C. Mavroforakis, A. Kyrillidis, and V. G. Vassiliadis. Improving co-cluster quality with application to product recommendations. In *Proc. of International Conference on Information and Knowledge Management (CIKM)*, pages 679–688, 2014. DOI: 10.1145/2661829.2661980 77

Y. Wakabayashi. Aggregation of binary relations: Algorithmic and polyhedral investigations. Ph.D. Thesis, University of Augsburg, Germany, 1986. 3

J. Wang, T. Kraska, M. J. Franklin, and J. Feng. Crowder: Crowdsourcing entity resolution. *Proc. of the VLDB Endowment (PVLDB)*, 5(11):1483–1494, 2012. DOI: 10.14778/2350229.2350263 99

D. P. Williamson and D. B. Shmoys. *The Design of Approximation Algorithms*. Cambridge University Press, 2011. DOI: 10.1017/cbo9780511921735 xiv

A. C.-C. Yao. Probabilistic computations: Toward a unified measure of complexity. In *Proc. of IEEE Symposium on Foundations of Computer Science (FOCS)*, pages 222–227, 1977. DOI: 10.1109/sfcs.1977.24 103

J. Yarkony, A. T. Ihler, and C. C. Fowlkes. Fast planar correlation clustering for image segmentation. In *Proc. of European Conference on Computer Vision (ECCV)*, pages 568–581, 2012. DOI: 10.1007/978-3-642-33783-3_41 21

C. Zahn, Jr. Approximating symmetric relations by equivalence relations. *Journal of the Society for Industrial and Applied Mathematics*, 12(4):840–847, 1964. DOI: 10.1137/0112071 2

H. Zha, X. He, C. H. Q. Ding, M. Gu, and H. D. Simon. Bipartite graph partitioning and data clustering. In *Proc. of International Conference on Information and Knowledge Management (CIKM)*, pages 25–32, 2001. DOI: 10.1145/502585.502591 77

Authors' Biographies

FRANCESCO BONCHI

Francesco Bonchi is Scientific Director at the ISI Foundation, Turin, Italy, where he's also co-ordinating the "Learning and Algorithms for Data Analytics" Research Area. Before becoming Scientific Director, he served as Deputy Director with responsibility over the Industrial Research area. Earlier, he was Director of Research at Yahoo Labs in Barcelona, Spain, where he led the Web Mining Research group. He is also (part-time) Research Director for Big Data & Data Science at Eurecat (Technological Center of Catalunya), Barcelona.

His recent research interests include algorithms and learning on graphs and complex networks (e.g., financial networks, social networks, brain networks), fair and explainable AI, and more in general, privacy and all ethical aspects of data analysis and AI. He has more than 200 publications in these areas. He also filed 16 U.S. patents, and got granted 9 U.S. patents.

He is member of the Steering Committee of ECML PKDD and IEEE DSAA, and is in the editorial board of several journals in the Data Science area. Dr. Bonchi has been the General Co-Chair of the 5th IEEE International Conference on Data Science and Advanced Analytics (DSAA 2018). He has been twice PC Co-Chair of the European Conference on Machine Learning and Principles and Practice of Knowledge Discovery in Databases (ECML PKDD 2010 and 2018), the 16th IEEE International Conference on Data Mining (ICDM 2016), the 28th ACM Conference on Hypertext and Hypermedia (HT 2017), the "Social Network Analysis and Graph algorithms for the Web" track at The International World Wide Web Conference (WWW 2018), and the 6th IEEE International Conference on Data Science and Advanced Analytics (DSAA 2019). Dr. Bonchi has also served as program co-chair of the first and second ACM SIGKDD International Workshop on Privacy, Security, and Trust in KDD (PinKDD 2007 and 2008), the 1st IEEE International Workshop on Privacy Aspects of Data Mining (PADM 2006), and the 4th International Workshop on Knowledge Discovery in Inductive Databases (KDID 2005). He is co-editor of the book *Privacy-Aware Knowledge Discovery: Novel Applications and New Techniques* published by Chapman & Hall/CRC Press. He will be General Chair of ECML PKDD 2023 to be held in Turin (Italy), and of ACM SIGKDD 2024, to be held in Barcelona (Spain).

DAVID GARCÍA-SORIANO

David García-Soriano is a Senior Research Scientist at the Institute for Scientific Interchange (ISI) in Turin, in the "Algorithmic Data Analytics" group. Previously, he received his Ph.D.

in Computer Science (2012) from the University of Amsterdam and his undergraduate degrees in Computer Science (2007) and Mathematics (2009) from the Complutense University of Madrid. He has been a member of the Algorithms and Complexity group at CWI Amsterdam (the Dutch National Research Center for Mathematics and Computer Science), and a research visitor at the Israel Institute of Technology in Haifa (Technion). Later, he was a postdoctoral researcher at Yahoo Labs Barcelona and a Lecturer in Computer Science at Pompeu Fabra University. He has also worked for industry as a software engineer at Google, CERN (the European Organization for Nuclear Research), and Tuenti. In recent years, he has been developing machine-learning and optimization-based solutions to financial portfolio management problems, in collaboration with Intesa San Paolo banking group.

His research focuses on the theory and practice of large-scale data mining and machine learning, with an emphasis on computational efficiency and provable quality guarantees; topics include algorithmic theory, combinatorial optimization, scalable machine learning, data mining, algorithmic fairness, social network analysis, data streams, and portfolio optimization. His research findings have been published in top-tier conferences (SODA, KDD, SIGMOD, ICALP, CCC, ICDM, WWW, ICDE, RANDOM, ECML/PKDD, SDM, …) and journals (SIAM Journal on Computing, Combinatorica, Data Mining and Knowledge Discovery, …).

FRANCESCO GULLO

Francesco Gullo is a *senior researcher* at the *UniCredit* banking group, specifically in the "Applied Research & Innovation" unit of the "AI, Data & Analytics ICT" department (UniCredit Services controlled company). Previously, he has been part of the "Research & Development" department (UniCredit holding company) for 5 years. He received his Ph.D., in "Computer and Systems Engineering," from the University of Calabria, Italy, in 2010. During his Ph.D., he was an intern at the George Mason University, U.S. After his graduation, he spent 1.5 years in the University of Calabria, Italy (as a postdoc), and 4 years in the Yahoo Labs, Spain (as a postdoc first, and as a research scientist then).

His research falls into the broad areas of *artificial intelligence* and *data science*, with special emphasis on *algorithmic* aspects. His recent interests include *mining and learning on graphs*, *natural language processing*, and *AI in finance*. He has been practicing both applied research (with a 10-year work experience in industrial-research environments), and fundamental research (with 80 publications in premier venues such as SIGMOD, VLDB, KDD, ICDM, CIKM, EDBT, WSDM, ECML-PKDD, SDM, TODS, TKDE, TKDD, MACH, DAMI, TNSE, JCSS, PR).

He has also been serving the scientific community, by, e.g., being Workshop Chair of ICDM'16, organizing workshops/symposia (MIDAS workshop @ECML-PKDD['16-'21], MultiClust symposium @SDM'14, MultiClust workshop @KDD'13, 3Clust workshop @PAKDD'12), or being part of the program committee of major AI/data-science conferences (e.g., SIGMOD, KDD, WWW, IJCAI, AAAI, CIKM, SIGIR, ICDM, WSDM, SDM, ECML-PKDD, ICWSM).

Printed in the United States
by Baker & Taylor Publisher Services